The Politics of Space

The Politics of Space

A History of U.S.-Soviet/Russian Competition and Cooperation in Space

Matthew J. Von Bencke

WestviewPress

A Division of HarperCollins*Publishers*

Copyright © 1997 by Westview Press, A Division of HarperCollins Publishers, Inc.

Published in 1997 in the United States of America by Westview Press, 5500 Central Avenue, Boulder, Colorado 80301-2877, and in the United Kingdom by Westview Press, 12 Hid's Copse Road, Cumnor Hill, Oxford OX2 9JJ

Library of Congress Cataloging-in-Publication Data
Von Bencke, Matthew J.
    The politics of space : A history of U.S.-Soviet/Russian competition and cooperation in space / Matthew J. Von Bencke.
        p.   cm.
    Includes bibliographical references and index.
    ISBN 0-8133-3192-7 (hardcover)
    1. Astronautics—International cooperation—Government policy—United States.   2. Astronautics—International cooperation—Government policy—Russia.   I. Title.
TL788.4.V65   1997
333.9'4—dc20   97-32067
                                                                                      CIP

The paper used in this publication meets the requirements of the American National Standard for Permanence of Paper for Printed Library Materials Z39.48-1984.

10   9   8   7   6   5   4   3   2   1

Contents

Figures

Preface

My interest in the coexistence of U.S.-Soviet/Russian competition and cooperation in space grew out of my interests in American and Soviet government, international relations and space policy. The natural overlap of these disciplines is American and Soviet/Russian space policy and how space policy relates to foreign and domestic policies. This disciplinary overlap provides the basis for a history of the space age which is focused on U.S.-Soviet/Russian international space policy formation. Rooting this history in its broader contexts relates space to wider issues concerning domestic and foreign policy formation in both the United States and the Soviet Union/Russia.

The initial stage of this project was made possible by funding from the Harvard Russian Research Center, the Public Broadcasting Service (PBS) and the Ford Program. This funding facilitated research trips to Washington and Moscow. Unfortunately my first trip to Russia was cut short when I was mugged and needed to return to the West for medical treatment. The rampant crime in Russia is quite real and threatens would-be researchers as well as economic recovery! The Ford Program and PBS were especially helpful in enabling me to return as quickly as possible to Moscow, where I continued my interviews and archival research.

The final stage of this project was made possible by the Center for International Security and Arms Control (CISAC) at Stanford University. CISAC and its co-directors, David Holloway and Michael May, provided me with a wonderful environment for research and writing, and without their financial, intellectual and personal support I could not have completed this book.

Special thanks go to Viktor Sokol'sky, a long-time Russian space historian, who served as my host and mentor at the Russian Academy of Sciences' Institute of the History of Science and Technology, as well as to Leonid Vedeshin of Interkosmos and Vladimir Kurt of the Institute of Space Research, both of whom commented on various aspects of my work and introduced me to several other Russian specialists who did the same. I am similarly grateful for the hospitality and assistance of Roger Launius, Lee Saegesser and Dill Hunley of the NASA History Office.

Many people were kind enough to comment on various aspects of my work. For such constructive advice I thank Graham Allison, David Bernstein, Roger Bourke, Ashton Carter, A. Denis Clift, Timothy Colton, Richard Davies, Paul Doty, Igor Drovenikov, Lewis Franklin, Arnold Frutkin, Marshall Goldman, Alexander Gurshtein, David Holloway, Frank Holtzmann, Nicholas Johnson, Nikolai Kardeshov, John Logsdon, Michael May, Andrei Piontkowsky, Robert Putnam, Vladimir Semenov, Vyacheslav Slysh, Marcia Smith, Lt. General Bernard Trainor (U.S. Marine Corps, ret.), Adam Ulam, Daniel Usikov, Nikolai Vorontsov, Celeste Wallander, Frank Winter, Victor Zaslavsky and Charlie Zraket. However, they are not responsible for or necessarily in agreement with the views expressed here, nor are they to be blamed for any errors of fact or interpretation.

Lastly, I am greatly indebted to three individuals. First, Joseph Nye was especially helpful in guiding the formative stages of this research. Dr. Nye encouraged me to pursue my interests in astrophysics, the Soviet Union *and* foreign policy, and helped guide the beginning stages of what has become a delightful amalgam of intellectual pursuits.

Second, Loren Graham taught me a deep respect for the study of the history of science, helped me arrange my research in Russia and provided valuable commentary on drafts of my work. His kind assistance was a great encouragement and boon to me throughout this project.

Lastly, I owe countless thanks to Anthony Oettinger. Dr. Oettinger consistently provided invaluable counsel, essential perspective and, when needed and when deserved, warm encouragement. He tirelessly read dozens of iterations of this work, and, in the process, not only gave me terrific advice, but also taught me how to be a better writer and scholar. Indeed, his magnanimous nature pervades the entire Program on Information Resources Policy at Harvard University which he chairs. I give my deepest gratitude and respect to him.

Matthew J. Von Bencke

Introduction:
International Space Policy--
A Paradigm for
Intergovernmental Relations

[Y]ears had passed. The storm-tossed sea of...history had sunk to rest upon its shores. The sea appeared to be calm; but the mysterious forces that move humanity (mysterious because the laws that govern their action are unknown to us) were still at work.

-- L. N. Tolstoy in *War and Peace*[1]

Tolstoy wrote this about Europe after the Napoleonic Wars, but many people might argue that the same passage applies to the years following the end of the Cold War. In the early 1990s, after nearly fifty years of uneasy coexistence under a nuclear cloud, it finally seemed that Russia and the United States might work together, and that peaceful cooperation might become the order of the day. An especially tell-tale sign of the improved international climate was the beginnings of U.S.-Russian cooperation in space; after all, space had historically been a favorite realm of superpower competition. However, a close look at U.S.-Soviet/Newly Independent States interactions in space suggests that it would be unreasonable to bet too heavily on the end of competition in space (and in other areas, for that matter). After all, weren't the same "mysterious forces" governing humanity even after the Cold War ended?

"What is the future of the world's two biggest space programs?" is an important question policy makers, participants and observers began asking themselves in earnest after the end of the Cold War; by the middle of the 1990s the debate was far from over. Nor has the fate of the United States' and the former Soviet Union's space programs been decided. The future continues to hang in the balance. Should the U.S. and Newly Independent States (NIS) work together in space? If so, in what ways, and how? What are the risks involved in such cooperation, and how can we minimize them? In what ways will competition continue to play a role? How will the private sector influence all of these issues?

Moreover, the answers to all of these questions depends upon and at the same time influences the outcome of much larger issues: what is the future of American relations with the Newly Independent States?

The middle of the 1990s is the ideal time to examine these and other related questions. Sitting at this juncture in history not only can we look at the American and Soviet space programs; we also have collected enough post-Cold War experiences to begin to extrapolate further into the future. In addition, perhaps most importantly, now is the time when this future will be determined. This book is an attempt to digest the history of the interaction of the U.S. and Soviet/NIS space programs, Cold War and post-Cold War, to fill the gap left by Walter McDougall's 1985 *The Heavens and the Earth.*[2] It is also a political history of the space age which, unlike its predecessors, focuses on the intersections of the world's two biggest space programs and the larger contexts within which these programs exist. Such a compilation is needed to inform the policy makers, participants and observers of the world's space programs, i.e., those who will influence the years to come.

The best way to peer into the future is to examine history. A quick glimpse at three seminal events in the space age suggests that the answers to all of these questions are complex. First, for example, when the Soviets launched the first artificial satellite, Sputnik, as a part of the International Geophysical Year (IGY), the newspapers of what became the two great space powers did not say, "Soviets Launch First Satellite: International Scientific Community Gains." Instead, the *New York Times'* headline declared, "Device is Eight Times Heavier Than One Planned by U.S."[3] *Pravda* talked not of international harmony but proudly exclaimed that "our contemporaries will witness how the freed and conscientious labor of the people of the new socialist society makes the most daring dreams of mankind a reality."[4] Thus, although it emerged in the context of an international year promoting the exploitation of space for all mankind, Sputnik was the starting gun to a space race largely characterized by superpower competition.

Second, partly in response to Sputnik, the United States, in July 1958, created the National Aeronautics and Space Administration (NASA). NASA was designed to successfully prosecute the space race, i.e., to win the competition with the Soviets. However, at the same time, the National Aeronautics and Space Act of 1958 encouraged the new space Administration to pursue "a program of international cooperation."[5]

Lastly, in 1960, Senator Kennedy declared that, "We are in a strategic space race with the Russians, and we have been losing....we cannot run second in this vital race. To insure peace and freedom, we must be first."[6] In May 1961 President Kennedy decided that the U.S. would undertake the greatest space challenge it could hope to achieve--sending men to the moon and back within the decade. However, although Kennedy embraced and accelerated the space race, he also asked in a September 1961 address to the United Nations,

> Why, therefore, should man's first flight to the moon be a matter of national competition?...Surely we should explore whether the scientists and astronauts of the two countries--indeed of all the world--cannot work together in the conquest of space, sending some day in this decade to the moon not the representatives of a single nation, but the representative of all our countries.[7]

Why this consistent duality? Were the American and Soviet International Geophysical Year efforts tailored towards sharing knowledge with other space scientists, or were they tools of an image contest? Was NASA supposed to achieve U.S. leadership in space or forge bonds with other nations? Did Kennedy want to cooperate or compete in space?

The answer to all of these questions is various shades of "both." In fact, throughout the space age, there has been a coexistence of U.S.-Soviet/Russian competition and cooperation in space. Space policy is the confluence of a number of interests, both domestic and international. When the confluences of American and Soviet/Russian motivations in space have overlapped, the two space powers have collaborated; when their collections of interests have dictated competition, the United States and the Soviet Union/Russia have competed. The appropriate question is not, "Did the U.S. and the USSR/Russia compete or cooperate in space in such-and-such a year?" but "To what degree and in what ways did the U.S. and USSR/Russia compete and collaborate during such-and-such a period?" The complicated calculations involved with this analysis reflect the fact that the Soviet and American governments are each multifaceted entities with a myriad of coexisting, often competing, interests. American and Soviet/Russian space policy is an outgrowth of this fact, and the American and Soviet/Russian space programs' parallel pursuit of competitive and cooperative policies is in many ways a paradigm for intergovernmental relations. At its essence, this work sets forth how the United States and the Soviet Union/Russia have historically competed and cooperated in space.

There are three additional factors that have served to focus the study of this history. First, the subject of this study is rooted in the larger context of overall American and Soviet/Russian domestic and foreign policy concerns. Similarly, although this work concentrates on those aspects of the U.S. and Soviet/Russian space programs which are related to interaction with other countries' space activities, it relates these aspects with the countries' overall space programs.

Second, space has civilian and military applications and implications. An important aspect of both superpowers' space programs has always been the interplay between their civilian and military space goals. This study takes into account both sides of the states' space programs.[8] However, a full analysis of the countries' military space programs requires a great deal of information that remains, in accordance with the U.S. government's thirty-year rule, classified as of the mid-1990s. Thus, with the exception of the early space age (from the mid-1950s to the early 1960s) this work's analysis gives greater emphasis to the American and Soviet/Russian civilian space programs than to their military space programs. Nonetheless, Chapter One presents a model for understanding the relationship between civilian and military space programs which has continued throughout the space age; the Conclusion takes advantage of and expands upon this model, examining the two major space powers' post-Cold War military space programs.

Third, space programs can be broadly divided into piloted and unpiloted efforts. Although I discus both areas of both the American and Soviet/Russian space programs, I emphasize the manned projects slightly more. This decision stems largely from the facts that (1) the piloted aspects of civilian space programs consume the most resources, and, as such, tend to be accurate indicators of the whole within which they exist, and (2) manned missions tend to capture both the public's and top-level politicians' attention more than unmanned missions; thus, since this work focuses on political aspects of the space programs, piloted projects are more germane. An exhaustive account of the coexistence of competition and cooperation in space would spend more time discussing the unmanned space programs, but its findings would not differ significantly from the conclusions reached here.

Chapter One, "The Origins of Space Policy: A Reflection and an Instrument of Foreign and Domestic Interests," describes the birth and development of the space age after World War II through 1958, examining how both Eisenhower's and Khrushchev's space policies were a reflection and an instrument of their foreign and domestic policy goals. Eisenhower and Khrushchev turned to military applications of rocket technology hoping to bolster their militaries while spending less on defense. Khrushchev, for his part, kept the Soviet military and civilian space programs intertwined and exploited the Soviet space program to bolster the Soviet Union's, the Communist Party's and his own personal image. In contrast, Eisenhower tried to keep the American civilian and military space programs as separate as possible. In addition, facts which surfaced in the early 1990s help show that Eisenhower made a conscious decision not to race the Soviets to space. Instead he sought to give an appearance of promoting space for peace while simultaneously paving the way for the U.S. military space program.

Sputnik's effects on U.S. and world opinion began an acceleration of the space race. Chapters Two, "The Context of Competition and Cooperation in Space: Overall U.S.-Soviet Relations and the American and Soviet Space Programs," and Three, "Early Days of Confrontation and Attempts at Cooperation," collectively cover the period from 1957 to 1969. These chapters (1) trace the early space age competition, (2) show the importance of the larger contexts in which cooperation in space occurred, i.e., the nations' overall relations and space programs and (3) discuss U.S.-Soviet interaction in the United Nations Committee on the Peaceful Uses of Outer Space (COPUOS). Despite their Cold War animosity, as the world's two major space powers, the United States and the Soviet Union sometimes had overlapping interests in the COPUOS.

Chapter Four, "Cooperation Come and Gone: The 1975 Apollo-Soyuz Test Project and the Subsequent Souring of Superpower Relations," covers the period from 1969 to 1980, highlighting the first joint U.S.-Soviet manned mission in space. Although the Apollo-Soyuz Test Project was largely a show of image, it marked the first significant confluence of American and Soviet interests in space. This period also demonstrated that warm overall U.S.-Soviet relations were a prerequisite for collaboration in space: the ebb and flow of cooperation from 1969 to 1980 closely matched the onset and downfall of détente.

Chapter Five, "A Confluence of U.S. and Soviet/Russian Interests in Space: Cooperation as the Cold War Ends," covers the period from 1981-1995, examining how the end of the Cold War led to markedly improved superpower relations and, correspondingly, to unprecedented levels of U.S.-Soviet (and then U.S.-NIS) cooperation in space.

Chapters Six and Seven, "Obstacles to Cooperation in Space, Post-Cold War," take a second look at the late 1980s and early 1990s, showing that the Cold War did not end up all roses for collaboration in space. Chapter Six, "A Nation in Disarray," focuses on developments inside Russia, the main successor to the Soviet Union. In the early 1990s the Russian civilian space program struggled under potentially crippling circumstances: Russia often seemed to be flirting dangerously with economic, social and political catastrophe, and the former Soviet civil space and science communities were fighting to maintain their infrastructure, to overcome organizational difficulties and to deal with inherited and largely anachronistic conservatism.

In addition, the dissolution of the USSR left the Soviet space program spread among newly sovereign nations, nations whose troublesome relations introduced new problems for the former Soviet space program. Chapter Seven, "International Barriers, New and Old," examines international challenges to post-Cold War cooperation in space, including relations between Russia and Ukraine and Russia and Kazakhstan. It also takes another look at an old international issue: the relationship between launch vehicles and weapons technology.

As the space age matured and commercial satellite applications expanded, private companies began to play a larger role in space. These developments, combined with the opening up of the former Soviet Union's space programs, paved the way for U.S.-NIS commercial space ventures. Chapter Eight, "The Market Bridges the Gap: Commercial Space Cooperation," examines and categorizes a number of U.S.-NIS non- or quasi-governmental collaborative space ventures, evaluating their motivations and their relation to intergovernmental space cooperation. This chapter also analyzes some of the difficulties these commercial interactions faced.

The Conclusion, "The Space Age Outlives the Cold War," examines how the end of the Cold War affected U.S. and Russian space policy. A part of this examination is a close look at the future roles of both nations' military space programs. These and related developments suggest that, despite the important changes in the world, the basic rules which guide space interaction outlived the Cold War.

As Chapter Seven and the Conclusion detail, non-Russian former Soviet republics inherited important aspects of the Soviet Union's space program. Thus, non-Russian Newly Independent States, especially Ukraine and Kazakhstan, are discussed. However, Russia dominated the NIS's space activities. As a result, and in pursuit of readability, post-Soviet space activities are referred to as "Russian," despite the fact that they often involve other former Soviet republics.

Now, on to delineating these "mysterious forces...."

Notes

1. Trans. Rosemary Edmonds (London: Penguin Books, 1978) 1339.

2. New York: Basic Books, 1985.

3. 5 Oct. 1957: A1.

4. 5 Oct. 1957. As cited in F. K. Krieger, *Behind the Sputniks* (Washington: Public Affairs Press, 1958) 311-12.

5. "National Aeronautics and Space Act of 1958" (PL 85-568, 29 Jul. 1958), *United States Statutes at Large*, vol. 72, pt. 1, 432.

6. John F. Kennedy, "If the Soviets Control Space, They Can Control the Earth," *Missiles and Rockets* 10 Oct. 1960: 12.

7. *Public Papers of the Presidents of the United States: John F. Kennedy, 1963* (Washington: Government Printing Office, hereafter GPO, 1964) 695. Hereafter referred to as *PP of JFK*.

8. This is not to suggest that the nations' civilian and military space programs are independent of one another; indeed, NASA, the Department of Defense, the Central Intelligence Agency and the National Reconnaissance Organization cooperate extensively, and all Soviet/Russian launches have been conducted by the Soviet/Russian military. Thus, as Chapter One will further delineate, the civilian and military space programs of both countries overlap.

1

The Origins of Space Policy: A Reflection and an Instrument of Foreign and Domestic Interests

Impressions created will color the thinking of the entire Russian military establishment and set the stage for future relations. Every...man therefore has in his hands...a profound responsibility. Our performance will be the yardstick by which the Russians judge the fighting capabilities, the discipline, the morale and the energy of the whole of the American forces....

> -- indoctrination addressed to the 15th U.S. Army Air Force on the morning of its departure for Soviet air bases; this uneasy cooperative mission entailed the first major deployment of a U.S. military force on Soviet soil, June 1994[1]

In World War II, the United States and the Soviet Union had fought the Axis Powers together. At the war's end this uneasy cooperative relationship degenerated until the two former allies openly vied for influence around the world, fostering alliances which would project and protect their power. There were clearly two Great Powers with two diametrically opposed ideologies and the world polarized around them. The United States, after conflicts over Eastern Europe, Berlin, Greece, Iran and Turkey, adopted a policy of encircling the Soviet Union with allies, trying to contain the presumably expansionist Communist bloc. The Soviet Union established dominion over its Central and Eastern European buffer zone, expanded its influence in Africa, China and the rest of Asia and even made forays into the Western Hemisphere.

The post-war situation was fertile ground for such bipolar competition.

9

The Great Powers of the Concert of Europe were largely no more. Great Britain and France saw their power diminish greatly; Germany was divided and occupied. Meanwhile the United States had evolved from a largely isolationist state to one that dominated the global economy, the possessor of an unscathed homeland and an image as the world's savior. The U.S. military was well supplied, stationed around the world and had suffered, relative to the other major powers' armed forces, few casualties. The Soviet Union, for its part, had a massive army that, like its American counterpart, occupied lands far beyond its borders, a land mass second to none and a claim as the leader of a massive Sino-Soviet Communist bloc that preached worldwide revolution.

As it turned out, the post-World War II U.S.-Soviet competition remained largely a war of nerves, a war of images. The points of contention were outside the two Great Powers' own territories. Confrontations arose and tensions ran high, but neither side was willing to engage the other in hot wars. Thus was born the Cold War. In recruiting their allies, both superpowers sought to portray themselves as the side to be on, the inevitable winner, the owner of the true ideology. Each country tried to, as was later expressed, win the hearts and minds of the Third World.

Sputnik

On Friday, October 4, 1957, the Soviet Union successfully launched *Sputnik*, a 184-pound satellite. This "moon" was the first artificial satellite to orbit the earth. The dramatic evidence of Soviet scientific success shocked the world--especially the United States public.

Like a tide raised by the "red moon," a wave of fear rumbled across the Western world. The Soviets had seized the initiative and started the space race, jumping from the starting gate, while the United States had yet to take its place at the starting line. As Khrushchev himself bragged,

> we were the first to launch rockets into space...first, ahead of the United States, England, and France. Our accomplishments and our obvious might had a sobering effect on the aggressive forces in the United States....They knew that they had lost their chance to strike at us with impunity.[2]

The United States had come to depend on technology as its greatest advantage in the superpower competition. Its atomic monopoly had ended in 1949, but its drive to maintain technological superiority to offset its inferiority in conventional land forces had been enshrined in NSC-162/2, Eisenhower's "New Look" strategy. NSC-162/2, officially titled "Basic National Security Policy," cited the "great Soviet military power" and the "basic Soviet hostility to the non-communist world, particularly to the United States," and resolved "to meet the Soviet threat" without "seriously weakening the U.S. economy."[3] In pursuit of these two goals the New Look policy sought to increase America's ability to counter the Soviet military while cutting defense spending by 30% over four years. The Eisenhower administration, simply stated, sought "more bang for the buck." Rather than maintain a huge army (and, in the event of hostilities, face an unacceptable number of casualties), America would use its know-how, its advanced nuclear weapons, rockets, planes, tanks, surface ships, submarines, radar, etc., to counter the hundreds of thousands of Communist troops. Thus, by beating America in what was presumably a race into space, the Soviet Union seemed to threaten the foundation of the security of the United States and its allies.

Furthermore, this technological public relations victory suggested something much more fundamental. Achievements in space were the luxuries of an advanced state which was wealthy enough to push back the frontier of science. The satellite was a product of the intellectual, engineering and manufacturing elite of the country. In sum, a better space program was, it seemed, a reflection of a superior society. The U.S. House Select Committee on Astronautics and Space Exploration provided a voice to the American public's shock, disappointment and guilt, explaining that "much of the responsibility" for America's failure to be first in space "must rest with the people as a whole, who spurned what warnings there were of the evolving Soviet menace, until they could hear the radio beep or see the Soviet sputniks as they passed overhead."[4] Had America become lazy and allowed its Golden Age to end already? Was Soviet Communism producing a finer education system, smarter scientists, better research and development and more efficient manufacturing? After all, was not this technological triumph merely a reflection of the fact that Soviet Communism was superior to American capitalism?

The Soviet Union, the vanguard of the world socialist movement, certainly thought so. In launching Sputnik, the USSR was making a statement to the world about its superior system of government. As *Pravda* claimed in announcing Sputnik's launch,

> Artificial earth satellites will pave the way to interplanetary travel and, apparently, our contemporaries will witness how the freed and conscientious labor of the people of the new socialist society makes the most daring dreams of mankind a reality.[5]

Seven years later in a similar article *Pravda* claimed that it was inevitable that the Soviets would be first in space: "This is natural...the so-called system of free enterprise is turning out to be powerless in competition with socialism in such a complex and modern area as space research."[6]

Space as a Reflection of Foreign and Domestic Policy

The Soviet Union

An examination of this, the dawn of the space age, reveals that for both superpowers the space program was an instrument and reflection of both domestic and foreign policy considerations. The interplay of these factors in the late 1950s and 1960s illustrates their coexistence throughout the space age.

The launching of Sputnik was a brilliant public relations coup, not a reflection of overall technological superiority. Sputnik's main achievement, since it was launched on a long-range military missile, was the highly public demonstration that the USSR possessed an advanced ICBM program; it also allowed the Soviets to test their tracking, power and transmissions systems. However, the satellite itself performed no specially designed experiments. Khrushchev had pushed for the early launch in order to make sure that it was the Communist standard bearer which ushered in the Space Age, ostensibly promoting the peaceful exploitation of space for the good of all mankind. The timing was perfect for such a projection, since the launch took place during and was advertised as a part of the International Geophysical Year (IGY), a year dedicated to initiating the peaceful use of outer space for all nations' benefit.[7]

However, the main motivation behind Sputnik's program was not exploiting space for the good of the world--it was exploiting space to prove that Soviet communism was the world's best system of government. The victory improved the Soviet image around the world and built upon two other recent highly publicized Soviet successes: four years before Sputnik's flight the Soviets announced the explosion of what they claimed was the world's first deliverable hydrogen bomb, and six weeks before Sputnik's launch the Soviets claimed possession of the world's first ICBM.[8] The Soviet space program had been ordered to launch the world's first satellite, and, as such, it was subject to foreign policy considerations. Thus, the Soviet space program was largely a reflection and an instrument of Soviet foreign policy.

Khrushchev also hoped to achieve military parity with the United States. This would be no small task: the Soviet economy was smaller than the U.S.'s, and the USSR had started this race behind, exploding its first atomic bomb more than four years after the United States had already dropped two. In addition, the Soviet population and infrastructure were heavily damaged during collectivization and the purges of the 1930s, not to mention during World War II, when the USSR suffered approximately 13,600,000 military and 7,720,000 civilian casualties (in contrast, the United States suffered 292,131 military and 5,662 civilian casualties).[9] By virtue of the fact that resources for the arms race came from the same budget which provided for domestic needs, this foreign policy goal was inextricably linked to Soviet domestic concerns.

One of the biggest problems Khrushchev faced in the mid-1950s was feeding the Soviet people. Soviet agriculture was stagnant--average farm yields did not increase from 1913 to 1953.[10] Famine loomed, and Khrushchev knew it. Compared to Western farming, Soviet agriculture was technologically backwards: Soviet farmers lacked the equipment, crop breeds, fertilizers and methods necessary to increase their efficiency. Self-proclaimed agricultural wizard Lysenko contributed to the problems by promoting outdated, untested and often backwards farming techniques. The looming disaster demanded rapid results, and this could only be achieved by dramatically increased investment.

However, this capital would not come easily. Khrushchev proposed to cut the military budget by one-third. In promoting this policy, Khrushchev embattled the huge Soviet military-industrial complex and all of its associated powers. The problem was, how could the Soviets compete successfully with American military power while spending less money on defense? How could Khrushchev placate the military-industrial complex?

The answer was a reliance on technology, on nuclear weapons. It was to these new weapons of mass destruction that Khrushchev, like his counterparts in the United States, turned for "more bang for the buck." The beauty of the logic must have seemed irresistible to a man in Khrushchev's position: a single nuclear weapon could wipe out an entire city or military base. Suddenly, instead of having to maintain enough divisions to be able to defeat the United States and its allies, one could argue that the Soviet Union needed only to build a nuclear arsenal capable of deterring, or, in the event of hostilities, destroying its enemies. A group of nuclear weapons would, in the long run, cost much less than the cost of maintaining a division and its support elements. Khrushchev openly acknowledged his reliance on nuclear weapons, stating in November 1959 that "[W]e now have stockpiled so many rockets, so many atomic and hydrogen warheads, that, if we were attacked, we would wipe from the face of the earth all of our probable opponents."[11]

All this meant more money for agriculture and the civilian economy. As Khrushchev reflected later in life,

Now that it's the size of our nuclear missile arsenal and not the size of our army that counts, I think the army should be reduced to an absolute minimum....When I led the Government and had final authority over the military allocations, our theoreticians calculated that we had the nuclear capacity to grind our enemies into dust, and since that time our nuclear capacity has been greatly intensified. During my leadership we accumulated enough weapons to destroy the principal cities of the United States, not to mention our potential enemies in Europe....If we try to compete with America in any but the most essential areas of military preparedness, we will be...exhausting our material resources without raising the living standard of our people. We must remember that the fewer people we have in the army, the more people we will have available for other, more productive kinds of work.... We must be prepared to strike back against our enemy, but we must also ask, 'Where is the end to this spiraling competition?'[12]

Still, developing the nuclear warheads and the missiles to deliver them was a gargantuan task. Khrushchev gave this research and development top priority; if he was to be able to save his country from agricultural disaster while competing with the United States militarily, the Soviet ICBM effort *had* to be successful. The research and development programs which were to launch Soviet satellites were the same as those which were to launch Soviet nuclear missiles; success in intercontinental rocketry was intimately tied with successes in space, and the increased emphasis on ICBMs served to accelerate the Soviet space program. In fact, Khrushchev allowed the satellite program to be born only as an offshoot of the military rocket program, once it had already proven successful.

Sputnik was launched on a modified *semyorka* missile, i.e., on the world's first ICBM (the NATO designation is "SS-6"). Khrushchev was hesitant to permit Korolev to launch a satellite because he feared it would delay the development of a deployable ICBM. Using a precious *semyorka* to launch a satellite would delay the missile program because a satellite launch could not provide information about warhead electronics or reentry. The Soviet government had begun officially considering launching a satellite in May 1954; however, the program's fate rested on Khrushchev's personal approval. Korolev did not gain that approval until January 1956, when Khrushchev was visiting the missile center, Kaliningrad. At the end of a day discussing ICBMs, Khrushchev, as he was preparing to leave, asked Korolev if there was anything else he would like to discuss. Korolev responded by asking for permission to use a *semyorka* to launch a satellite. Khrushchev paused, and Korolev hurriedly interjected that "the Americans were already working on the launch of an artificial satellite." He claimed that the USSR could launch not only before the United States, but that it could launch a satellite several times bigger than the America's.

The opportunity to give the Americans a public "punch in the nose" enlivened Khrushchev, and he began asking Korolev how much the *semyorka* would have to be modified to launch a satellite. Korolev responded, "All we will do is remove the thermonuclear warhead and replace it with a satellite. That is all...."

Khrushchev responded, "If the main goal will not suffer, do it."[13] Many years later Khrushchev would remember this comment that became a seminal event in the history of the space age, writing,

> When we became convinced that Korolev had solved the problem of designing a rocket for space exploration, we were able to step into the international political arena and show that now even the territory of the United States of America was vulnerable to a strike by our missile forces. It became, as they say, a balm to the soul, and our position had improved.[14]

The Soviet Space Program and Khrushchev's Rise to Power

Khrushchev also used the space program to his own personal benefit. Khrushchev's position as the top Soviet leader was hard won and anything but stable. When Stalin died in March 1953, he left no clear successor. Although the Politburo declared active the principle of "collective leadership" in the wake of Stalin's death, Malenkov, Molotov, Kaganovich, Beria and Khrushchev all vied to increase their own personal power vis a vis the others.'[15] Few doubted that one leader would emerge preeminent in this system so tailored towards one-man rule; it simply remained to be seen who that person was.

Few people thought that man would be Nikita Khrushchev. In fact, in March 1953 the Politburo listed him as fifth in power.[16] Nonetheless, by 1957, after helping to subordinate the powerful state security apparatus to political leaders, taking advantage of the nomenklatura system to appoint loyal officials, arranging for the demotion of Malenkov and a reconciliation with Tito's Yugoslavia and delivering his remarkable anti-Stalin speech at the Twentieth Communist Party Congress in October 1956, Khrushchev stood at the pinnacle of Soviet power. However, this pinnacle was not stable; in June 1957 the Politburo demanded Khrushchev's resignation from his post as First Secretary, and it was only with the help of the military (as led by Marshal Zhukov) that Khrushchev survived this attack on his power. As a part of his struggle to power and of his effort to bolster his personal image as the Soviet Union's best-fit leader, Khrushchev made himself the space program's, and especially the cosmonauts,' personal patron.[17] In addition to directly intervening in rocketry research and development to make sure it received the resources it needed, Khrushchev arranged press conferences, parades and press accounts to portray himself as the genius behind Soviet space achievements. So it was, for example, with national fanfare that Khrushchev met and subsequently paraded Yuri Gagarin, the world's first man in space. It was only due to Khrushchev's personal interventions, Soviet citizens were to think, that the Soviets were so successful in space. This was a critical aspect of the creation of Khrushchev's public image--indeed, of his own personality cult. He used this image to consolidate and increase his own power, or, in short, to improve his claim to being *the* leader of the Soviet Union. These were the Soviet Union's greatest public relations achievements, and it was Khrushchev who positioned himself to personally reap the image benefits.[18]

Khrushchev also inherited a greatly weakened Party. Stalin had risen above the Party and had worked around it. Under Stalin, Party Congresses were held extremely infrequently (for example, thirteen years elapsed between the Eighteenth and Nineteenth Congresses), the Politburo had been sub-divided (thereby increasing dissension and decreasing efficiency) and Party leaders had been arbitrarily removed, persecuted and executed. Leading politicians were subject to accusations, trials and punishments administered by a security apparatus subordinate only to Stalin. So weakened was the Party in 1953 that Malenkov chose to pursue the pinnacle of power by remaining Chairman of the Council of Ministers while yielding his position as First Party Secretary. It was Khrushchev who capitalized on this mistake, skillfully using his powers as First Secretary to his advantage.

In order to remain the Soviet Union's leader, Khrushchev had to strengthen the Party. By reinvigorating the Party, the First Secretary stood to gain in terms of personal power; indeed, for him there was no other way. Reinvigorating the Party also meant furthering (or, at the least, *appearing* to further) its ultimate victory, i.e., the advancement of the world socialist movement.

How could Khrushchev revitalize the Party and its revolutionary fervor? One way was to use propaganda, and space--that mysterious, dreamy, seemingly unconquerable void--was the perfect raw material for Khrushchev's propaganda mills. Khrushchev improved the Party's standing by declaring achievements in space to be the products of the Party-led socialist movement. Soviet successes in space, Khrushchev declared, were a product of the labor of all socialist people, a part of the advance of socialism towards achieving full knowledge and surpassing the doomed bourgeois imperialists.

Thus, the early Soviet space program was an instrument and reflection of both domestic and foreign policy goals. The Soviet space program was tailored largely to achieve dramatic image boosts. At the same time, in seeking to increase its global power reach by striving towards parity while avoiding an agricultural crisis, the USSR placed great emphasis on its rocket programs. Lastly, Khrushchev personally used the space program to bolster his and the Party's strength.

The United States

The relative openness of the American government facilitates a close-up examination of the formation of early U.S. space policy. American leaders, like their Soviet counterparts, sought to reduce their military budgets while increasing their military power largely through developing intercontinental ballistic missiles. However, unlike Khrushchev, Eisenhower resolved to keep the American military and civil rocket programs as separate as possible.

Rather than promote the most promising route for a quick satellite launch, the American leaders acted explicitly according to other, more diverse priorities. The United States, if its leaders had pushed for it, could have "beaten" Sputnik into orbit. The technology was there, but the priority was not. Keeping this in mind, early American space policy can be understood by analyzing three related priorities. First, its military missile programs (which were given top priority as a part of the New Look strategy) and civilian rocket development were kept as separate as possible. Second, in organizing the fledgling space program, the Eisenhower administration sought to project the U.S. as the leader in promoting space for peace. Third, the U.S. military and intelligence establishment was trying to lay the foundations of its future military space program. Investigating the manner in which the Eisenhower administration sought to realize these three priorities reveals that the early U.S. space program, like the Soviet program, was a product of foreign and domestic policy considerations.

What Race?

These non-race-oriented priorities reveal that the stark terms of the importance of image in the space race had, before Sputnik, not yet become clear to the American leadership. Though America's image turned out to be a driving force behind American's space program from its beginning, it is, ironically, that very image which suffered when it failed to launch first.

In 1957 America's leaders learned that not only was space to be an important part of the superpowers' international image competition, but that the space race would be, like the Cold War, total. There was no top-down push within the United States government to launch the world's first satellite. This decision resulted not so much from a dismissal of Soviet capabilities or from a confidence in America's "lead" as much as from a failure to recognize the value of launching a satellite first. President Eisenhower never pushed for the U.S. to be the first to do so; in fact, he insisted accurately five days after the Sputnik launch that the American "satellite program has never been conducted as a race with other nations."[19] Eisenhower, who played an active role in defense and related decisions, had specifically decided not to "race" the Soviet Union to space.

Reactions to Sputnik

It is useful to closely examine the context of Eisenhower's decision. The charged context of the Cold War competition highlights the fact that in deciding not to race the Soviet Union to space, the Eisenhower administration was placing higher priorities elsewhere. Moreover, this examination describes how with Sputnik the superpower image competition entered a new, more intense phase.

It is not that there were no predictions regarding the public relations impact the first satellite launch would have; in fact there were several such predictions which reached the highest levels of the executive branch. The stark and often perspicacious nature of the warnings highlights the importance the administration gave its other priorities.

First, to examine the warnings themselves. As early as 1946 a RAND study predicted that the first satellite launch "would inflame the imagination of mankind, and would probably produce repercussions in the world comparable to the explosion of the atomic bomb."[20] Similarly, NSC Directive 5520 warned that the Soviets were working on their own satellite program and noted that

> Considerable prestige and psychological benefits will accrue to the nation which first is successful in launching a satellite. The inference of such a demonstration of advanced technology and its unmistakable relationship to intercontinental ballistic missiles technology might have important repercussions on the political determination of free world countries to resist Communist threats, especially if the USSR were to be the first to establish a satellite.[21]

The top of the executive branch was aware of these potential repercussions. A classified report requested by President Truman and presented to President Eisenhower upon its completion predicted that a satellite

> would be considered of utmost value by the members of the Soviet Politburea....the satellite would have the enormous advantage of influencing the minds of millions of people the world over during the so-called period

of 'cold war....' it should not be excluded that the Politbureau might like to take the *lead* in the development of a satellite. They may also decide to dispense with a lot of the complicated instrumentation that we would consider necessary to put into our satellite to accomplish the main purpose, namely, of putting a visible satellite into the heavens first. If the Soviet Union should accomplish this ahead of us it would be a serious blow to the technical and engineering prestige of America the world over. It would be used by Soviet propaganda for all it is worth.[22]

As the previous discussion of Sputnik's simplicity and Soviet propaganda shows, Dr. Grosse could hardly have been more correct. Also, in an April 27, 1955 letter to Dr. Alan Waterman, the Director of the National Science Foundation (which was overseeing the U.S. involvement in the IGY), Deputy Under Secretary of State Robert Murphy noted that the U.S. satellite project, if successful, would "undoubtedly add to the scientific prestige of the United States, and it would have a considerable propaganda value in the cold war."[23]

These predictions were largely borne out after Sputnik's launch. The world, entrenched in the Cold War and still gripped by the tension of the crises in Hungary, the Middle East and Poland, was carefully watching the superpower competition.[24] Domestic and world opinions were sensitive to what seemed to be the eclipsing of the leader of the West--in its supposed forte. American scientists dining with their Soviet counterparts in observation of the International Geophysical Year that October evening "were caught completely by surprise."[25] *The New Republic* declared that the launch was proof that "the Soviet Union has gained a commanding lead in certain vital sectors of the race for world scientific and technological supremacy,"[26] while the *New York Times* bemoaned the apparent relative feebleness of American plans with the headline "Device Is Eight Times Heavier Than One Planned by U.S."[27] *Newsweek*, in seeking to answer the "most chilling" question, "What effect would the Soviet achievement have on the nation's security?" noted that in the future "sputniks would be able to sight and even photograph just about every point on earth," and, most ominously, worried that the "successful launching of the satellite gave strong support to Russia's claim that it has a workable intercontinental ballistic missile."[28]

Politicians, reflecting and fueling the growing public hysteria, joined the chorus of the concerned. The White House was "engulfed in turmoil."[29] Senator Henry Jackson (D - Washington) called Sputnik "a devastating blow to the prestige of the United States as the leader in the scientific and technical world," the Democratic Advisory Council (which included Adlai Stevenson and former President Truman) concluded that "[t]he all-out effort of the Soviets to establish themselves as master of space [sic] around us must be met by all-out efforts of our own," and even President Eisenhower admitted that the Soviets had "gained a great psychological advantage through the world."[30]   Senator John Kennedy (D - Massachusetts), who would soon, as President, initiate the century's grandest space venture, emphasized the resulting global loss of prestige:

> It seems to me, and this is most dangerous for all of us, that we are in danger of losing the respect of the people of the world....[T]he people of the world respect achievement. For most of the 20th century they admired American science and American education, which was second to none. But now they are not at all certain about which way the future lies. The first vehicle in outer space was called sputnik, not Vanguard.[31]

Senator Lyndon Johnson (D - Texas) led those demanding to know if, how and why the United States had allowed its security to go lax and initiated the "Preparedness Investigating Subcommittee of the Committee on Armed Services." This subcommittee produced some 3290 pages of testimony, all because

> [o]ur country is disturbed over the tremendous military and scientific achievement of Russia. Our people have believed that in the field of scientific weapons and in technology and science that we were well ahead of Russia. With the launching of Sputniks I and II [*Sputnik II* was launched on November 3]...our supremacy and even our equality has been challenged.[32]

Indeed, from the public's point of view, American security and the technical supremacy upon which this security depended had been dramatically, convincingly challenged.[33]   The Red nemesis' leader had declared "We will bury you," and the eery "beep, beep, beep" of Sputnik (as publicly broadcasted in the U.S.) echoed Khrushchev's threat all too loudly.

Fears that the United States was losing a scientific and technological race went beyond the White House meeting rooms. One national survey found that the "majority" of Americans agreed that "Sputnik means the Russians have beaten us scientifically."[34]   A congressional report concluded that "unless this country makes a larger effort, in the next ten to twenty years, it will find itself definitely left behind....the United States faces a very real possibility of being outdistanced on the scientific front in an age when science in application can decide the fate of nations."[35]

The American image suffered abroad, too. One survey showed that the majority of the citizens of New Delhi, Toronto, Paris, Oslo, Helsinki and Copenhagen agreed that Sputnik "struck a hard blow at U.S. prestige."[36] Such findings confirmed American leaders' fears that the disturbingly successful Soviet campaign to appear to be the world's scientific leader would scare allies into neutrality and neutral nations into the Eastern Bloc.

American politicians could not afford to stand by idly. Sputnik forced a firm response. After Sputnik it was beyond question that the space race was total. Not only did image dominate the space programs, making them instruments and reflections of foreign policy, but the resulting image was dominated by appearing to be the best--first. As the NASA Administrator who oversaw the Apollo era wrote, "The great issue of this age is whether the U.S. can, within the framework of the existing economic, social, and political institutions, organize its development and use of advanced technology as effectively for its goals as can the Soviet Union...."[37]

"Total" Space Race

Given this damage to the American image, why did the U.S. not try to launch its satellite as soon as possible? Part of the answer to this question lies in the advantage of hindsight. Eisenhower did not take the actions required to lead the United States into space first partly because he did not realize the public relations power of being first in space. Even after the Soviet scientific and public relations coup, Eisenhower declared, "I can't understand why the American people have got so worked up over this thing."[38] "Eisenhower was genuinely puzzled by the panic over Sputnik....[A]ccording to James Killian [Special Assistant to the President for Science and Technology], the President had no idea that the American public was 'so psychologically vulnerable.'"[39]

In addition, part of Eisenhower's inaction may be attributed to the overabundance of information and viewpoints and the pace with which the actual events unfolded. However, the most important factor to consider in understanding this crucial phase of the early American space program is that Eisenhower's decision to not push for as early a satellite launch as possible was mostly a reflection of different priorities.

Space for Peace

The significance of Eisenhower's decision not to prioritize as early a satellite launch as possible is highlighted by the fact that the United States, had speed been a priority, could have been the first nation to orbit an artificial satellite. The warnings and the technology existed, but the priorities were elsewhere. One of the priorities deemed more important than speed was the presentation of an image as promoting "space for peace," and this motivation helps explain why the different military space programs were handled the way they were.

Each military service had its own satellite program. The Army pursued and promoted its satellite program, which was run by the Army Ballistic Missile Agency (ABMA). Wernher von Braun, the leading German rocket scientist who fled, along with most of his colleagues, to the United States at the end of World War II, led the Army program. The ABMA proposed to use a modified Jupiter missile, the Jupiter-C, to launch a satellite. Meanwhile, the United States Air Force (USAF), asserting its dominion over what it defined as the "aerospace" continuum, presented its Thor and Atlas boosters as best suited to lead America into space (the Air Force was already in charge of the highly classified WS [Weapon System] 117L project, which designed and built America's first reconnaissance satellites; further discussion follows beginning on page 25). Lastly, the Navy had been researching satellites since 1945, and its Viking rocket was the centerpiece of its space program as run by the Naval Research Laboratory (NRL). Each of the three service's space organizations sought survival and growth at the expense of the others. The Army, Air Force and Navy--especially in the context of Eisenhower's budget cuts--each vied for the honor (and accompanying funding) which they assumed would come with the first satellite launch. Indeed, the survival of the ABMA, Air Force and NRL projects depended upon their ability to prove their unique worth.

These projects existed parallel to each other for several years, but, as the IGY drew closer, one of them had to be selected to provide the rocket for America's first satellite launch. Assistant Secretary of Defense for Research and Development Donald A. Quarles led the decision making process. Quarles initiated the advisory Stewart Committee (chaired by Homer Stewart of the Jet Propulsion Laboratory), which considered each service's proposal. The resulting decision came in a vote on August 3, 1955. Three committee members voted for the Navy proposal (Project Vanguard), two for the Army's (Project Orbiter).

This close vote, which determined that the Navy would be responsible for the United States' first satellite launch, illustrates the heated nature of the debate and at the same time reflects the administration's priorities. RAND and Quarles had told the committee members that the administration sought a space program with as strong a civilian character as possible. Eisenhower's public relations priority was to portray the American space program in terms of "space for peace." In terms of this criterion, the decision was an easy one: the Viking rocket was developed by private industry, whereas the Jupiter rocket was produced by an Army arsenal which was directed by von Braun, a former Nazi scientist whose V-2 rockets had damaged Allied targets during World War II. It is in view of these factors that Eisenhower decided "to entrust the admirals with the job."[40]

It is ironic that this public relations priority contributed to the decision which facilitated America's "losing" the satellite race, a loss which created a veritable public relations debacle. Vanguard not only failed to beat the Soviets into space, but it also failed to beat the ABMA--even though the Army project, after the Stewart Committee decision, was only marginally supported.[41] Clear evidence--much of which was available at the time of the decision--demonstrates that the ABMA could have launched a satellite sooner than the Navy.[42] In fact, recently declassified documents make it clear that the officials making the decision knew that, but simply did not prioritize speed. In revisiting the controversial decision the Department of Defense's Special Assistant for Guided Missiles, E. Murphree, wrote in a memorandum,

> I have looked further into the matter of the use of the Jupiter re-entry test vehicle as a possible satellite vehicle in order to obtain an earlier satellite capability as we discussed recently. I find that there is no question but that one attempt with a relatively small effort could be made in January 1957.[43]

In another memorandum Murphree wrote

> While it is true that the Vanguard group does not expect to make its first satellite attempt before August 1957, whereas a satellite attempt could be made by the Army Ballistic Missile Agency as early as January 1957, little would be gained by making such an early satellite attempt.[44]

Moreover, decision makers believed a January 1957 launch would not be as valuable as a later launch, both scientifically and politically:

> such a single flight would not fulfill the Nation's commitment for the International Geophysical Year because it would have to be made before the beginning of that period. Adequate tracking and observation equipment for the scientific utilization of results would not be available at this time.[45]

However, the administration's emphasis on a civilian program in pursuit of a positive image could prove effective only so long as the United States was the first to launch a satellite. The damage to America's image, domestically and internationally, which resulted from the Soviets beating the United States into space was, as the preceding descriptions of the hysteria and disillusionment indicate, greater than that which may have been caused by the near-subtlety of who produced the rocket which launched the world's first (*American*) satellite. The variation in military/civilian involvement between Project Vanguard and Project Orbiter was one of gradations: both projects were, in the end, controlled by the Navy and the Army, respectively.

Instead of stressing an early launch, the American satellite effort was emphasized as one aimed towards launching a satellite during the IGY which could provide useful data to scientists around the world. In a presidential statement of July 29, 1955 Eisenhower first announced that he had

> approved a plan by this country for going ahead with the launching of small unmanned Earth-circling satellites as part of the United States participation in the International Geophysical Year....the American program will provide scientists of all nations this important and unique opportunity for the advancement of science.[46]

As the President would point out after Sputnik's launch, the United States fulfilled these goals. From the start, the American space program emphasized long-term benefits from using satellites with advanced microelectronics in order to perform military and civilian functions including reconnaissance, communications, meteorology, navigation and measurements of the composition and properties of space. These priorities and America's strong microelectronics industry both resulted partly from and contributed to each other. *Explorer 1*'s tiny 10.5-pound package included micrometeoroid detectors, Geiger counters and telemetry equipment, and discovered the Van Allen radiation belts. All of the relevant data was presented openly to scientists through the IGY. This, the Americans fairly claimed, was the exploitation of space for the benefit of all mankind. Thus, the Vanguard civilian satellite program was favored to appear to be promoting space for peace and to keep the military and civilian rocket programs as separate as possible.[47]

Military Versus Civilian Rockets

There was, however, more at work than "space for peace." An October 9, 1957 White House statement accurately describes an aspect of America's early rocket programs' priorities, claiming that "Merging of this...effort [to launch an American scientific satellite] with military programs could have produced an orbiting United States satellite before now, but to the detriment of scientific goals and military progress." The desire to keep both the military and civilian rockets programs and to keep them apart was derived partly out of a desire to present the U.S. civilian space program as one pursuing space for peace, and partly out of the practical desire to keep the ICBM program on track:

> It would be impossible for the ABMA group to make any satellite attempt that has a reasonable chance of success without diversion of the efforts of their top-flight scientific personnel from the main course of the Jupiter

program, and to some extent, diversion of missiles from the early phase of the re-entry test program.

The United States ICBM program had, as of 1953, been roughly four years behind the Soviets'.[48] American security demanded that this gap be closed if not reversed; Eisenhower's hope to offset cuts in conventional forces with superior strategic forces magnified this need.

Space for Power: Paving the Way for American Eyes in the Sky

There was a third reason why the United States did not race to space. One of the primary uses American leaders saw for space was reconnaissance. The nature of the Cold War required military planners to prepare their forces to respond quickly to threats the world around. This required quick, effective, accurate intelligence--globally. This need was all the more urgent given that the Soviet Union was a closed, highly secretive society. The U.S. knew that the Soviets were developing new long-range bombers, overseas bases, submarine bases and perhaps most importantly ICBMs; the questions were where they were, when, and what were they doing. As Eisenhower stated his concern, "Modern weapons had made it easier for a hostile nation with a closed society to attack in secrecy and thus gain an advantage denied to the nation with an open society."[49]

Thus, even before Sputnik was launched, the United States gave a high priority to improving intelligence gathering and early warning capabilities. The American intelligence community was, in the early 1950s, desperate for information about the Soviet Union, and did its best with outdated German aerial reconnaissance photographs from the early 1940s, interrogations of returned German and Japanese prisoners of war and reconnaissance flights along Soviet borders. In pursuit of better sources of data the Air Force, with Eisenhower's approval, released 516 camera-equipped balloons from Western Europe in January and February 1956. However, the project was only marginally successful, as it depended on the balloons chancing to drift over useful, unobscured targets during the day.[50]

In order to improve upon the balloons the Air Force turned to high altitude piloted aircraft. The U-2 reconnaissance plane was built by Lockheed's California "skunk works" and could fly at over 70,000 feet (no fighter known of then could operate at more than 50,000 feet). In late 1955 the U-2 became, at Eisenhower's direction, primarily the Central Intelligence Agency's project,[51] and, in June 1956, the U-2 made its first flight over the Soviet Union. As was the case with Project Genetrix, the

U-2 overflights had Eisenhower's explicit approval. The U-2 was a dramatic improvement over the wandering balloons. Still, the strip of land a U-2 could cover was relatively limited. Furthermore, American leaders suspected that the Soviets would eventually develop effective countermeasures, and saw the U-2 as a stopgap used in anticipation of even more effective high altitude reconnaissance vehicles.[52]

The more effective vehicle they had in mind was the military satellite. Satellites could provide pictures more effectively than any means previously available, covering broad swaths of the earth extremely rapidly. Improvements in optics promised highly detailed images. Although satellites could not fly around clouds like planes often could, they were unstoppable in the sense that nothing could shoot them down. Furthermore, high-altitude reconnaissance could, in the event of hostilities, be an effective force multiplier, directing the appropriate forces to respond most efficiently at specific points; satellite meteorology could augment this multiplier.

Thus, American officials, beginning in studies of the mid-1940s, saw the space program as a future replacement for and improvement upon balloon and plane espionage. As early as 1945, interrogations of German World War II rocket scientists suggested the utility of space espionage.[53] An October 4, 1950 RAND study predicted that satellites could provide secure communications, "strategic and meteorological reconnaissance of the kind that...would be of high military value" and act as relay stations for guiding long-range missiles.[54] In March 1954 RAND predicted that a spy satellite system would produce 30 million pictures annually--a number equivalent to all the pictures the USAF had acquired from all sources in the previous twenty-five years,[55] and recommended that the "Air Force undertake the earliest possible completion and use of an efficient satellite reconnaissance vehicle" as a matter of "vital strategic interest to the United States."[56]

That same spring of 1954 Eisenhower, as a strategic thinker and former Supreme Allied Commander ever fearful of another Pearl Harbor-like surprise attack,[57] appointed a Surprise Attack Panel of leading scientists. The Surprise Attack Panel, which was renamed the Technological Capabilities Panel, had three subcommittees, one on offensive forces, one on defensive forces and one on intelligence, and was intended to propose ways to avoid surprise attack. James R. Killian, president of MIT from 1948 to 1959, chaired the Panel, and the intelligence subcommittee was led by Edward Purcell, a Nobel Prize-winning Harvard physicist, and Edwin Land, the inventor of the Polaroid camera, head of the Polaroid Corporation and a personal advisor to every American president from 1955 to 1970. On February 14, 1955 Killian and Land briefed Eisenhower on the Panel's final report, which concluded that

We must find ways to increase the number of hard facts upon which our intelligence estimates are based, to provide better strategic warning, to minimize surprise in the kind of attack, and to reduce the danger of gross overestimation or gross underestimation of the threat. To this end, we recommend the adoption of a vigorous program for the extensive use, in many intelligence procedures, of the most advanced knowledge in science and technology.

The form of this "most advanced knowledge" was omitted from the printed report for security reasons, but, as Killian and Land orally conveyed to the President, it was reconnaissance satellites.[58] Killian and Land urged Eisenhower to improve high altitude espionage capabilities, starting with balloons and airplanes and leading to satellites.[59] The historical development of Project Genetrix, the U-2 and Project Corona[60] show that Eisenhower followed their advice precisely.

Further research broadened the functions of military satellites. In the summer of 1958 NSC Directive 5814/1 predicted that military satellites would produce revolutionary advances in espionage, communications, weather forecasting, electronic countermeasures and navigation.[61]

Thus, from its beginning, the Air Force's 117L classified military space program received a high priority. The USAF asked RAND to study the issue, and in April 1951 RAND issued a report confirming the potential utility of spy satellites. RAND then subcontracted the issue to the Beacon Hill Study Group, which was chaired by Carl Overhage of Eastman Kodak. In June 1952 the Beacon Hill Study Group issued its final report, in which it concluded that improvements in electronics and optics could provide useful intelligence from high altitudes. In response the Air Force issued contracts to Radio Corporation of America, North American Aviation, Westinghouse Electric Corporation, Bendix Aviation, Allis-Chalmers and the Vitro Corporation to begin research and development of satellite television cameras, vehicle guidance equipment, attitude-control devices and nuclear auxiliary electrical power sources. On March 1, 1954 this private consortium, dubbed Project Feedback, concluded a common report detailing the hardware requirements for observation, cartographic and weather satellites and their ground support systems. The Air Force, riding the wave created by the Technology Capabilities Panel, awarded the first contract for spy satellite development to Lockheed on October 29, 1956, initiating what became Weapon System 117L. Project Vanguard had not even begun, but the road was thoroughly paved for America's military space program.[62]

However this road was not seen as without obstructions. One of the problems American planners saw with developing the necessary fleet of spy satellites was that the Soviets were sure to respond hostilely. Despite the American position that sovereignty over space did extend to space itself, the Soviets, they believed, were sure to object that American satellites were violating their sovereign air space.[63] In order to set a legal precedent for satellite overflight, the United States decided first to orbit as innocuous a satellite as possible. As NSC Directive 5520 said, "a small scientific satellite will provide a test of the principle of 'Freedom of Space.' The implications of this principle are being studied within the Executive Branch. However, preliminary studies indicate that there is no obstacle under international law to the launching of such a satellite."[64] A related RAND study titled, "The Satellite Rocket Vehicle: Political and Psychological Problems," concluded

> It is very doubtful, however, whether the USSR would accept such a 'vertical' limitation of sovereignty; and in any case, it would certainly not regard the passage of any vehicle in the outer space over its territory as 'innocent' if it were demonstrated that it was being used to perform acts which in themselves infringe upon sovereignty. We may assume that satellite operations designed to gather visual information in Soviet territory, if they become known to the Soviet leaders, will be construed by them as a 'consummated act of aggression.' This reaction is likely to verbalized in legalistic terms.[65]

In order to test the waters of what the Soviet leadership would consider "a major threat," the study concluded that "best policy seems to be to stress the experimental nature of instruments in, and communications with" a "preliminary" satellite. Then a "second 'work' satellite" "to be used for intelligence purposes" could be more safely launched.[66]

Early preparations for the satellite program show that this was in fact the course taken. The program, though intended to appear as civilian as possible, was initiated by Quarles at the Department of Defense.[67] Once under way, the International Geophysical Year satellite program was handled as a classified matter in close cooperation with the Department of Defense and the Central Intelligence Agency. In a classified May 1955 letter Dr. Waterman, the Director of the National Science Foundation, updated Quarles on the civilian project's progress. Dr. Waterman noted that he had "discussed the subject with Allen Dulles [Director, CIA], with Richard Bissell present, the latter being the one in Central Intelligence who is following this closely" and that "Dulles volunteered to present the subject to the Operations Coordinating Board in order to get action started." In addition, Waterman stated, the subject was "thoroughly canvassed" at the Department of State, and the Bureau of the Budget had "informally agreed as to the importance of the matter and their cooperation when action is needed. They agreed to keep the matter confidential...."[68]   According to Waterman, the Department of State's approval was sought only after the plan had been drawn up with Quarles' guidance.[69] *Explorer 1* did more than introduce the United States to space, it specifically paved the way for the American military space program.

Meanwhile Quarles, who had become the Secretary of the Air Force, made sure that the military satellites came second, slowing down the Air Force's reconnaissance satellite construction in November 1956, less than a month after the USAF had issued the first development contract to Lockheed.[70] The military satellite would come, but its birth was carefully controlled to follow that of its civilian cousin.

Thus, the early American space program was arranged in such a way as to present itself as promoting space for peace, to keep it as separate as possible from the development of ballistic missiles and to set a precedent for the military uses of space. The foreign policy considerations inherent in these international image and national security considerations were intertwined with Eisenhower's desire to derive more military might for less money.

America Awakened

The launch of Sputnik and the ensuing American reaction are seminal events in Cold War history. The shock of the Soviet launch, avoidable though it might have been, caused a heightened awareness among American leaders that space was crucial to the superpower image competition. Moreover, this Soviet propaganda victory elevated this image battle to a higher level, forcing the Americans to struggle to appear to "catch up." In this struggle the Eisenhower administration finally changed its original plans and allowed the Army to launch a satellite before Project Vanguard.

Not since Pearl Harbor had the American consciousness been so galvanized into action. Eight days after Sputnik's launch Eisenhower conferred with the Department of Defense's Office of Defense Mobilization's Science Advisory Committee (SAC) in what turned out to be a critical meeting. The SAC advisors (including Land and Killian) convinced Eisenhower that while the United States was still ahead of the Soviet Union in science and technology, it was rapidly losing this lead. Science was given a much higher priority in Russian culture, and their schools were a powerful reflection and guarantor of this fact. Unless the trend was reversed, Land argued, "Russian scientific culture will leave us behind as a decadent race."[71] Extrapolating from the current trends, the SAC concluded that in ten years the Soviet Union would be technologically superior. This meeting had a dramatic effect on

Eisenhower, who referred to it repeatedly in the weeks to come.[72]

Eisenhower and the Congress enacted broad changes in response to these warnings. These changes fit into three categories: education, governmental research and development organization and the American civil space program. First, in answer to the perceived crisis in education, on September 2, 1958 the President signed into law the National Defense Education Act (NDEA). Before this Act, American education had almost entirely been the responsibility of state and local governments. The NDEA established an unprecedented role for the federal government in education, authorizing just under $1 billion in primary and secondary education aid, a $295 million college loan fund which gave special consideration to students planning to teach or specialize in science, math or foreign languages, $280 million to help schools purchase laboratory equipment and an additional $59.4 million for graduate school students specializing in areas related to national defense.[73]

Second, Eisenhower reorganized the Department of Defense, strengthening the position of the Department's Secretary and trying to diminish the ill effects of interservice competition. Under the new arrangement the Secretary of Defense gained control of all operational commands and the entire Department of Defense budget, which he then dispersed to the individual services. The Joint Chiefs of Staff (JCS) was also strengthened and recast to be a cooperative body with an integrated staff designed to direct all armed forces in peace and in war. These measures were devised largely in order to increase the efficiency of research and development within the overall armed forces.

In addition, the Department of Defense's Science Advisory Committee was elevated to the White House, where it became the President's Science Advisory Committee (PSAC). Killian chaired this body as Eisenhower's first Special Assistant for Science and Technology from 1957 to 1959. This elevation was a part of the overall greater emphasis given to research and development, on which total government spending increased steadily from $4,462 million in 1957 to $4,991, $5,806 and $7,744 million in 1958, 1959 and 1960, respectively.[74]

Congress also acted to support and oversee future science and space policies. The Senate created the Committee on Aeronautical and Space Sciences and the House created the Committee on Science and Astronautics.

Lastly, on July 28, 1958, Eisenhower signed into law the National Aeronautics and Space Act of 1958, creating the civilian National Aeronautics and Space Administration (NASA).[75] NASA was expressly designed to make America the world's leader in space technology and exploration. The new Administration subsumed NACA (the National Advisory Committee for Aeronautics) and also absorbed Project

Vanguard, the Jet Propulsion Laboratory and part of the Army Ballistic Missile Agency. The civilian agency was to exercise "control over aeronautical and space activities sponsored by the United States, except [those]...primarily associated with the development of weapons systems, military operations, or the defense of the United States;" these responsibilities remained with the Department of Defense.[76]

With the new government involvement in education (especially scientific education), the streamlined federal research and development and Department of Defense structures and the new National Aeronautics and Space Administration Eisenhower formed the basis for America's participation in the newly begun space age. The United States would not stand by while the Soviets pressed their gains. As Eisenhower reflected in January 1958,

> what makes the Soviet threat unique in history is its all-inclusiveness. Every human activity is pressed into service as a weapon of expansion. Trade, economic development, military power, arts, science, education, the whole world of ideas--all are harnessed to this same chariot of expansion. The Soviets are, in short, waging *total cold war*.[77]

In the wake of Sputnik, America joined this battle, and the fight for world leadership entered a new realm in earnest.

Notes

1. Thomas A. Julian, "Operations at the Margin: Soviet Bases and Shuttle-Bombing," *The Journal of Military History* Oct. 1993: 640.

2. Nikita Sergeevich Khrushchev, *Khrushchev Remembers*, trans. Strobe Talbott (Boston: Little, Brown and Company, 1970) 516-17.

3. NSC 162/2 was the second version of report number 162 of the National Security Council (NSC), an executive advisory board which assists the president in the formation of national security policy. The National Security Council was created on July 25, 1947 as a part of the National Security Act. NSC 162/2 was classified until 1977, with access to it "very strictly limited on an absolute need-to-know basis," and was circulated to the NSC members, i.e., Secretary of the Treasury, Attorney General, Director of the Bureau of the Budget, Chairman of the Council of Economic Advisors, Chairman of the Atomic Energy Commission, the Federal Civil Defense Administrator, the Chairman of the Joint Chiefs of Staff and the Director of the Central Intelligence Agency. This report called for the development and maintenance of "a strong military posture, with emphasis on the capability of inflicting massive retaliatory damage by offensive striking power," an allied rapid deployment capability, a mobilization base "adequate to insure victory in the event of general war," a broad intelligence system and the maintenance of a "sound, strong and growing economy...[and] morale and free institutions." The report noted and predicted the continuance of U.S. and Soviet

reluctance to initiate general warfare; the atomic "stalemate" it described became the Cold War, and NSC 162/2 is largely the birth of American Cold War policy. President Eisenhower directed "all appropriate executive departments and agencies" to implement it immediately upon its dissemination October 30, 1953. Paul Kesaris, *Documents of the National Security Council 1947-1977* (Washington: University Publications of America, 1980) (reel 3, #1062, microfilm).

4. *The National Space Program* 85th Congress, 2nd Session (H. Rpt. 1758, *Series Set 12073*) (Washington: GPO, 1958) 30.

5. *Pravda* 5 Oct. 1957.   As cited in F. K. Krieger, *Behind the Sputniks* (Washington: Public Affairs Press, 1958) 311-12.

6. *Pravda*, "Sorry, Apollo!" 11 Oct. 1964.   Apollo was the American manned moon program; Chapter Three, pages 66-70. As cited in James E. Oberg, *Red Star in Orbit* (New York: Random House, 1981) 77.

7. The IGY was organized by the World Committee for the International Geophysical Year (CSAGI).   In a 1954 Rome meeting CSAGI recommended that "thought be given to the launching of small satellite vehicles, to their scientific instrumentation, and to the new problems associated with satellite experiments, such as power supply, telemetering, and orientation of the vehicle." The IGY started on July 1, 1957 and ended on December 31, 1958.

8. The Soviet hydrogen bomb, called Joe-4, was exploded August 12, 1953, nine months after the United States exploded the world's first (non-deliverable) hydrogen bomb.   Joe-4 (named after Joseph Stalin) was the product of a rapid catch-up effort and was the first sign that the Soviet Union was capable of competing with the United States technologically.

On August 27, 1957 the USSR announced possession of an ICBM after successful tests of a *semyorka* (R-7, or "ol' number seven"); this announcement was treated with a certain amount of skepticism by the United States public, but was given new credence by Sputnik's launch (page 19, note 28).   The *semyorka* was designed chiefly by Sergei Korolev, the Soviet Union's leading rocket scientist.

9. Chris Cook and John Stevenson, *Longman Handbook of World History Since 1914* (New York: Longman, 1991) 50.   The numbers for the Soviet Union are estimates.

10. Roy A. Medvedev and Zhores A. Medvedev, *Khrushchev: The Years in Power* (New York: W. W. Norton and Company, 1978): 27.   Soviet crop yields were one-third those of other European countries in 1953.

11. Ervin Jerome Rokke, "Politics of Aerial Reconnaissance: Eisenhower Administration," diss., Harvard University, 1971, 125.   Even as Khrushchev publicized his reliance on nuclear weapons, he greatly exaggerated his country's nuclear forces. Chapter Two, pages 45-46.

12. Khrushchev, *Khrushchev Remembers* 517-19.

13. Sergei Khrushchev, *Krizisi i Rakety* (Crises and Rockets), vol. 1 (Moscow: Novosti, 1994) 110-11.   Sergei Khrushchev was the son of Nikita Khrushchev and accompanied his father on this and many other trips.

14. Nikita Sergeevich Khrushchev, *Khrushchev Remembers: The Glasnost Tapes* (Boston: Little, Brown and Company, 1990) 187.

15. Beria, the head of the Ministry of Internal Security (MVD), was the first to lose in this competition when Malenkov, Molotov, Kaganovich and Khrushchev cooperated in arranging his arrest and subsequent execution four months after Stalin's death. In so doing the other four leaders began the weakening of the internal security apparatus, which had, under Stalin, ruthlessly exploited its power, including its ability to charge, arrest, try and punish even regional Party chairs. The MVD was subsequently split into the MVD and Committee for State Security (KGB).

16. Medvedev and Medvedev, *Years in Power* 4.

17. In so doing Khrushchev borrowed a page from Stalin's notebook: Stalin was the self-described "father" of Soviet aviation.

18. It is worth noting that meanwhile, Korolev, the true genius behind the early Soviet space program, was kept in relative obscurity, where he would die, frustrated.

19. "Statement by the President Summarizing Facts in the Development of an Earth Satellite by the United States" (9 Oct. 1957) 735. In *Public Papers of the Presidents of the United States: Dwight D. Eisenhower, 1957* (Washington: GPO, 1958) 733-735. The statement goes on to say, "The United States satellite program has been designed from its inception for maximum results in scientific research. The scheduling of this program has been described to and closely coordinated with International Geophysical Year scientists of all countries." Hereafter referred to as *PP of DDE*.

20. Douglas Aircraft Company, Inc., "Preliminary Design of an Experimental World-Circling Spaceship," Rpt. no. SM-11827, 2 May 1946, 17. Referred to in the Works Consulted as NASA History Office reference no. 4 (NHO 4). This study was commissioned at the request of Major General Curtis E. LeMay, the first Deputy Chief of Air Staff for Research and Development of the Army Air Forces, and was first presented May 15, 1946 at a meeting of the War Department's Aeronautical Board, the coordinating board for early American satellite research and development. Rand was initiated as Project RAND by the Army Air Forces commander, General H. H. Arnold, as an independent consultant group. R. Cargill Hall, "The Origins of U.S. Space Policy: Eisenhower, Open Skies, and Freedom of Space," *Colloquy* Dec. 1993: 5.

21. NSC Directive 5520, 20 May 1955. This Directive was rescinded on January 28, 1960. NHO 26.

22. Aristid V. Grosse, "Report on the Present Status of the Satellite Problem," 25 Aug. 1953. NHO 11. Dr. Grosse was President of the Research Institute of Temple University, was involved with the Manhattan and Vanguard Projects and wrote this report in consultation with Dr. von Braun (then Director of the Guided Missile Development Group of the U.S. Army Ordinance at Huntsville, Alabama), the Pentagon, Naval Research Laboratory, Wright Field Air Force Base, Rand Corporation and Columbia University. Dr. Wallace H. Graham, Truman's personal physician, arranged for the report's preparation. Dr. John R. Dunning of Columbia University, a close scientific advisor to Eisenhower, personally discussed the report with Eisenhower in late 1953/early 1954. Letter from Dr.

Grosse to Dr. Eugene Emme, 12 Jan. 1973. NHO 12. This report urged that steps be taken to rapidly expand the U.S. satellite effort. Later Dr. Graham said that Truman lamented not taking full advantage of Dr. Grosse's "knowledge and foresight," saying American projects "would have been far [more] advanced" had his advice been heeded. In this letter to Dr. Grosse, Dr. Graham reports, "It is perhaps I was [sic] not able to present this in proper fashion as it was thwarted by the Navy at that time and they gave the President the information that they had plans well in hand." Letter from Dr. Graham to Dr. Grosse, 17 Jan. 1973. NHO 10. Despite Dr. Graham's willingness to assume culpability, there were other, larger forces at work--i.e., higher executive priorities--as will be discussed in the following pages.

23. Robert Murphy, "Confidential Memorandum for Dr. Alan T. Waterman." NHO 23.

24. Appendix B, the Chronology of the Space Age, notes more specifically the Hungarian Revolution, Soviet-Polish confrontation and 1956 Middle Eastern war. This Chronology lists important political and international events as well as all piloted and important unpiloted American and Soviet/Russian space missions through early 1994; it is intended to help the reader see how the U.S. and Soviet/Russian space programs have been heavily influenced by larger world events.

25. Robert A. Divine, *The Sputnik Challenge* (New York: Oxford UP, 1993) xiii.

26. "Sputniks and Budgets," 14 Oct. 1957: 3-4.

27. *The New York Times*, 5 Oct. 1957: A1.

28. *Newsweek*, "Satellites and Our Safety," 21 Oct. 1957: 29.

29. Divine, *Sputnik Challenge* 3.

30. *Newsweek* 21 Oct 1957: 29. Divine, *Sputnik Challenge* 7.

31. U.S. Senate. Committee on Commerce, Subcommittee on Communications. *The Speeches, Remarks, Press Conferences, and Statements of Senator John F. Kennedy, August 1 through November 7, 1960* 87th Congress, 1st session (S. Rpt. 994, part 1) (Washington: GPO, 1961) 159. Speech given on September 7, 1960, in Portland, Oregon.

32. U.S. Senate. Committee on Armed Services, Preparedness Investigating Subcommittee. *Inquiry Into Satellite and Missile Programs* 85th Congress, 1st Session (Washington: GPO, 1957) 1-2. Although all of the politicians mentioned (with the exception of the President) were Democrats, regardless of whatever their partisan motivations may or may not have been in bemoaning the American mistakes, the fact relevant here is that they did seize upon, reflect and feed the public hysteria which followed Sputnik.

33. For example, a national survey revealed that the "majority" of Americans agreed that if "Russia can shoot up a satellite, its missile program must also be ahead of ours." An "overwhelming majority" agreed that the Russian missile program was a direct threat. "The U.S., Ike, and Sputnik" *Newsweek* 28 Oct. 1957: 35.

34. *Newsweek*, "The U.S., Ike, and Sputnik," 32.

35. U.S. House. Select Committee on Astronautics and Space Exploration. *The National Space Program* (H. Rpt. 1758) 29.

36. "Six out of Seven," *Newsweek* 28 Oct. 1957: 35.

37. James E. Webb, *Space Age Management* (New York: McGraw-Hill, 1969) 382.

38. Eisenhower said this on October 15, 1957. Divine, *Sputnik Challenge* 12.

39. Divine, *Sputnik Challenge* 17.

40. Letter from Dr. Grosse to Dr. Eugene Emme, 12 Jan. 1973. NHO 12. Quote is John Dunning's paraphrase of Eisenhower's words. Note 19 (page 19) contains the official explanation.

41. Project Orbiter was kept alive by General Medaris with low-level funding without Pentagon knowledge or approval. Divine, *Sputnik Challenge* 9.

42. The army claimed that it could launch a satellite by January 1957. The army's intermediate-range ballistic missile Jupiter-C rockets were based on the already established Redstone rockets, whereas the Viking rockets were more complex and untested. The months following the Stewart Committee decision evidence this conclusion: on September 20, 1956 a Jupiter-C missile flew three thousand miles, reaching a height of 600 miles. On August 8, 1957 another Jupiter-C was successfully launched--and its nose cone became the first man-made object ever recovered from space. After September 1956, the ABMA rocket "had a 50% or better chance of placing a satellite into orbit." Walter A. McDougall, *The Heavens and the Earth* (New York: Basic Books, 1985) 131.

Even more compelling is the fact that these achievements were reached *after* Project Orbiter had been rejected (it had minor funding and no Pentagon support). Later Assistant Secretary Quarles admitted that if Project Orbiter had been approved in August 1955, the United States could have launched a satellite a year before Sputnik. Divine 10. As it was, in the wake of Sputnik, the army was given the go-ahead to launch a satellite--which it did on January 31, 1958 by orbiting *Explorer 1*. Thus, even after 29 months of relative obscurity, the army beat the navy to space (the navy finally launched *Vanguard 1* on March 17, 1958).

43. E. Murphree, "Memorandum to Reuben Robertson re Jupiter," 5 Jul. 1956. Declassified 25 May 1994. NHO 21. Mr. Robertson was Assistant Secretary of Defense for Research and Development (after Quarles).

44. "Use of the Jupiter Re-entry Test Vehicle as a Satellite," 5 Jul. 1956. Declassified 25 May 1994. NHO 22.

45. Homer J. Stewart, "Memorandum for the Assistant Secretary of Defense (R&D)," 22 Jun. 1956. Declassified 25 May 1994. NHO 29. Mr. Stewart was Chairman of the Advisory Group on Special Activities.

46. Statement by James C. Hagerty, The White House, 29 Jul. 1955. U.S. Senate. Committee on Aeronautical and Space Sciences. *Documents on International Aspects of Exploration and Use of Space 1954-1962* 88th Congress, 1st Session (Document No. 18, *Serial Set* 12555) (Washington: GPO, 1963) 27.

47. It is worth recalling at this point that the American civil and military space programs were heavily intertwined. For example, the Mercury program (Chapter Two, page 51 and Chapter Three, page 67) used the Redstone and Atlas boosters; the Gemini program (Chapter Three, page 69) used the Titan booster. Each of these three boosters was developed, produced and operated by the Department of Defense.

48. McDougall, *Heavens* 99.

49. Hall, *Origins* 19.

50. This project was called Weapon System 119L, or Project Genetrix, and ran from November 1955 to February 1956. Earlier balloon projects were even less successful. Stephen E. Ambrose, *Ike's Spies: Eisenhower and the Espionage Establishment* (Garden City: Doubleday, 1981) 268.

51. Paul B. Stares, *The Militarization of Space: U.S. Policy, 1945-1984* (Ithaca: Cornell UP, 1985) 32.

52. The leaders' concerns were finally confirmed when, on May 1, 1960, Gary Powers' U-2 plane was shot down over the Soviet Union. Hall, *Origins* 6, 21, 23.

53. A spy 'rocket,' it was predicted in 1945, "could complete one circuit around the earth in any time between one and one-half hours and several days. The whole of the earth's surface could be continuously observed from such a rocket....and be able to observe even small objects, such as ships, icebergs, troop movements, constructional work, etc." Dr. F. Zwicky, "Report on Certain Phases of War Research in Germany," 1 Oct. 1945, vol. 1. NHO 36. Dr. Zwicky was the Director of Research at Aerojet Engineering Corporation and a Professor of Astrophysics at California Institute of Technology. The report was classified.

54. Paul Kecskemeti, "The Satellite Vehicle: Political and Psychological Problems," Rand memorandum RM-567, 4 Oct. 1950: 2, 3 and 5.

55. Hall, *Origins* 19.

56. Stares, *Militarization* 31.

57. Ambrose, *Spies* 267.

58. Jeffrey T. Richelson, *America's Secret Eyes in Space* (New York: Harper and Row, 1990) 12.

59. Richelson, *Eyes* 12-13.

60. Project Corona became the Central Intelligence Agency's (CIA) first reconnaissance satellite program and was initiated when on August 18, 1960 the CIA recovered the first satellite photographs of the Soviet Union from the *Discoverer 14* satellite.

61. August 1958. For further discussion of this important NSC directive, see Chapter Two, page 43.

62. Stares, *Militarization* 30-31. Hall, *Origins* 6, 19.

63. The Soviets did indeed protest American reconnaissance satellites. Chapter Two, pages 44-45 and 55-56.

64. 20 May 1955: 3. NHO 26.

65. Kecskemeti, Rand RM-567: 15.

66. Ibid., pp 13, 11, 22 and v.

67. In a classified March 18, 1955 memorandum Dr. Waterman informed Robert Murphy that the "United States National Committee for the International Geophysical Year has been considering, *at the suggestion of the Assistant Secretary of Defense for Research and Development,* the feasibility and scientific importance of inclusion in the United States program of the launching of a small satellite for geophysical (*unclassified*) purposes during the International Geophysical Year" (italics added). Declassified November 15, 1989. NHO 32.

68. Letter from Dr. Waterman to the Honorable Donald A. Quarles, 13 May 1955. Declassified November 15, 1989. NHO 33.

69. Letter from Dr. Waterman to Mrs. Constance Green, 15 Jul. 1965. NHO 34. As these letters show, Waterman, though ostensibly the head of the American civil satellite program, was well aware of his role as the harbinger of American military uses of space.

70. Hall, *Origins* 22.

71. Divine, *Sputnik Challenge* 13.

72. Divine, *Sputnik Challenge* 12-15.

73. "National Defense Education Act of 1958" (PL 85-864, 1 Sep. 1958), *United States Statutes at Large*, vol. 72, pt. 1, 1580.

74. McDougall, *Heavens* Appendix.

75. "National Aeronautics and Space Act of 1958" (PL 85-568, 29 Jul. 1958), *United States Statutes at Large*, vol. 72, pt. 1, 426.

76. As the previous discussions of the American IGY satellite program and military satellite program show, the United States has, from their beginning, tried to keep the civil and military space programs separate. This clause of the National Aeronautics Act of 1958 codified this dichotomy, but, as the intimate relationship between the armed forces, CIA and the NSF and the subsequent interaction between NASA and the Department of Defense show, the lines between the American civil and military programs have always been hazy and flexible.

77. *PP of DDE, 1958* (Washington: GPO, 1959) 9. Annual Message to Congress on the State of the Union.

2

Early Days of
Confrontation and
Attempts at Cooperation,
1957-1962

Sputnik was only the beginning of the race in space. After the October 1957 launch the race's pace quickened. The Soviet Union and the United States pushed their scientific and engineering resources to produce results which would launch satellites, and thereby evidence their social, economic, political, scientific and ideological superiority. After Sputnik the Soviets launched *Sputnik 2*. It was then America's turn to enter space. American space officials got an early taste of the dangers of public failure on December 6, 1957 when the navy's *Vanguard TV-3* satellite burned on the launch pad. The event had been highly publicized as the United States' first attempt to orbit a satellite, and was heavily criticized as "Kaputnik," "Stayputnik" and "Flopnik." Finally, on January 31, 1958, the Army Ballistic Missile Agency successfully orbited America's first satellite, *Explorer 1*. The Navy followed on March 17 with *Vanguard 1*. In the next twenty-one months the Soviets launched *Sputnik 3*, *Luna 1* (the first spacecraft to escape Earth's gravity entirely), *Luna 2* and *Luna 3*, while the Americans orbited *Vanguard 2*, *Pioneer 4* and *Explorer 6*. By the end of 1960, the Soviet Union had thirteen successful launches; the United States had thirty-three (and also thirty-four failures; the corresponding Soviet number is unknown).[1] The Soviets touted their payload's great weights and their firsts, while the Americans boasted of their spacecraft's scientific uses.

Thus continued the total space race. However, even as the competition of the space race intensified, the founding principles and possibilities of cooperation were explored. The account of the first part of the space age, from 1945 to 1958 (Chapter One), illustrates the interplay between domestic and foreign policy influences in the formation of both Soviet and American space policy, and, thanks to the availability of declassified information for that era, is especially useful in gaining an understanding of the interplay between the American civil and military space programs. This chapter and Chapter Three examine the second part of the space age, covering events from 1957 to 1969. This latter period is especially useful for studying space law, the role of international organizations (particularly the United Nations) and the connection between the superpower image battle and the related (and now expanding) understanding of the interplay between the two nations' space programs' competitive and cooperative elements. First, it is necessary to investigate how the United Nations became involved in the space race, and how the United States and Soviet Union utilized and interacted within the United Nations.

International Implications of the Space Age: Space Law

Sputnik introduced heretofore unheard of legal problems. The natural forum for consideration of many of these questions was the United Nations, and it was in this same forum that states' representatives strove to promote cooperative exploitation of space for the benefit of all mankind. In the process, the United Nations not only laid the foundations of space law and of the principles of space cooperation, but it also intensified and provided a forum for U.S.-Soviet competition to appear as the state pursuing space for peace.

Thus, competition coexisted with cooperation. The Americans and Soviets tacitly cooperated, as, in spite of their adversarial relationship, the two great space powers had common interests in preserving their space programs' independence from outside controls. Neither power wanted to freely cede their investments to other countries, yet both had to face other nations' opinions, and both stood to gain from some degree of international support (both materially, in terms of host countries for tracking facilities, and immaterially, in terms of public international support).

Thus, scientific one-upmanship (and the resulting practical benefits) was only the first aspect of the early space race. A second goal of the space powers was to purvey the best public image, abroad and domestically. Third, the U.S. and the Soviet Union sought to pursue these goals while promoting the foundations of space law which best advanced their individual (and sometimes common individual) interests and capabilities. The interplay of these three factors further explains early Soviet and American civilian space policy.[2]

The Need for Space Law

The launch of Sputnik forced the world's nations, and especially the world's two great space powers, to debate space law more earnestly. The world's first satellite transformed hypothetical legal quandaries into very real debates. Satellites were a technological breakthrough; they could pass over dozens of countries in a matter of hours. It was not clear if this was a violation of the passed-overs' sovereign rights. Other legal questions concerned the allocation of radio frequencies, potential contamination from space, access to celestial bodies, the registration of space vehicles, who would benefit from the exploitation of outer space, what form international regulation of the exploration of outer space would take, how liability would be defined, whether or not countries had an obligation to return astronauts and their equipment after crash landings and what kind--if any--of military uses of space were legal. Underlying all of these issues (as is true of all international law) was the question of who would decide them.

Precedents

The Paris Air Convention of 1919 and the Chicago Convention on International Civil Aviation of 1944 gave states sovereignty over the *air space* above their territories.[3] One problem posed by the space age was the exact definition of "air space." Did "air space" have an upward limit? If so, what might that limit be?

Although the Soviet Union boycotted both conventions,[4] they actively supported the Chicago Convention's definition of the principle of national sovereignty. The USSR considered its air space to be just as much its territory as the land itself.[5] They used this principle to criticize American high altitude reconnaissance balloons as illegal (recall Weapon System 119L and its predecessors). In contrast, Secretary of State Dulles defended the balloons' legality, claiming they flew above the limits of national sovereignty.[6]

Sputnik led to a shift in Soviet policy on how far sovereignty of air space extends. Apparently in preparation for Sputnik's launch, this shift began in March 1956. A Soviet legal scholar who explained official state policy justified Sputnik as perfectly legal for two main reasons. First, he said, it was a part of the International Geophysical Year, and, as such, was part of an international activity. Sovereignty, the scholar argued, does not extend past "air space" and into "outer space," and, besides, he declared, Sputnik does not pass over territories--the territories revolve under Sputnik.[7]

This change in Soviet position was welcomed by the United States. Since 1950 American policy makers were concerned about establishing the legality of satellites.[8] Until the Soviets justified their launch of Sputnik, United States officials were unsure what the ramifications of launching a satellite might have been.

Thus the debate about whether or not satellites violated air space sovereignty was settled quietly by tacit agreement. Eventually a functional definition of "space" was accepted: objects are considered to be in space when they are in orbit. This tacit cooperation enabled the Americans and Soviets to launch satellites without having to worry about overflight permissions--to an extent.

Other issues were not so easily resolved. In addition to the United States' and Soviet Union's differences, other countries wanted to join the fray. The forum for these debates became the United Nations.

The United Nations

The Soviet bloc and the United States submitted rival plans to create an ad hoc UN committee to deal with space matters. On November 24, 1958 the American plan prevailed, forming the United Nations Ad Hoc Committee on the Peaceful Uses of Outer Space (COPUOS). The plan passed by a vote of 54-9-18; all nine votes against were cast by socialist states. This was a significant victory for the United States and presented an international impression of the U.S. as the leading proponent of space for peace and the benefit of all peoples.

The Committee was significant both for what it was intended to do and for what it was not intended to do. American views dominated. It was generally agreed that the ad hoc COPUOS would form working groups to study space law. Furthermore, the Committee was to explore cooperation possibilities, information exchanges, the division of radio frequencies and spacecraft registration and liability. The Committee was *not* intended to be a precedent for a United Nations space agency (as the Soviets had hoped)[9] and would not immediately undertake space disarmament negotiations.

The Soviet Union boycotted the Committee in protest of its composition and its membership, which heavily favored the Western voting bloc.[10] The Soviet Academy of Sciences formed its own Commission on the Legal Problems of Outer Space, and its chairman, E. A. Korovin explained his country's position by declaring that

> [t]he U.S. delegation...used its 'voting machine' to carry a resolution on the creation of such a rigged preparatory group (two-thirds of whose members are tied to U.S. military blocs) as would give the U.S. complete control of it. Thus, the U.S. Government has again showed its intention to subordinate cosmic research to its expansionist and aggressive plans....This is why the Soviet Union naturally refused to participate in this body.[11]

The Soviet boycott reinforced the United States' ability to portray itself as leading the fight for space for peace, as American plans were accepted and passed without serious opposition.

In July 1959 the Ad Hoc Committee finished its first report, which strongly supported American policy. In fact, this report closely resembles the National Security Council's study, "Preliminary U.S. Policy on Outer Space."[12] The survey emphasized the benefits mankind stood to gain from the exploration and exploitation of space (e.g., communications, weather forecasting, cartography and navigation), declared space free for all peoples' usage, mentioned no definition of where space "begins," and did not attempt to create an international space agency nor to negotiate a demilitarization of space. Instead, the report focused on the United Nation's ability to pursue modest managerial functions, such as registering spacecraft and dividing up the radio spectrum.

Avoiding the Extension of National Rivalries?

After issuing the summer 1959 report, the United Nations worked to make the COPUOS permanent. It did so on December 12 of the same year in Resolution 1472 (XIV), "International Co-operation [sic] in the Peaceful Uses of Outer Space." The resolution attached "great importance to a broad development of international co-operation in the peaceful uses of outer space in the interests of the development of science and the improvement of the well-being of peoples,"

> believing that the exploration and use of outer space should be only for the betterment of mankind and to the benefit of States irrespective of the stage of their economic or scientific development [and] desiring to avoid the extension of present national rivalries into this new field.[13]

It called for a review of possible areas of cooperation, especially as a part of the International Geophysical Year, for a study of relevant legal problems and a conference for the exchange of information. The resolution was hardly proactive, and thus fulfilled the U.S.'s goal of being able to circumvent possible restrictions. This ability to avoid controls benefitted both the United States' and the Soviet Union's space programs.

Still, the resolution included an important concession to non-space powers, that all peoples should benefit from the usage of space, *irrespective of their stage of economic or scientific development.* This was a concession which would come to haunt the space powers.[14]

Neither the United States nor the Soviet Union wanted to freely share their space technology. To do so would reduce their incentive to invest in active space programs. Still, the United Nations' members only included two superpowers and many other voices. Non-space powers saw possible future benefits, feared being left out in the cold and wanted protection against exploitation by the space powers.[15]

Resolution 1472 also added six members to the Committee on the Peaceful Uses of Outer Space, and, in response, the Soviets discontinued their boycott of the Committee.[16] However, the end of the Soviet boycott only brought national rivalries into COPUOS meetings in the form of stifling debate. Resolution 1472 was passed unanimously, giving the now-permanent Committee a strong endorsement. However, the COPUOS was paralyzed for two years by haggling over procedural issues, such as officer designations, voting procedures and the mechanics of the world scientific conference planned for 1961.[17] The Soviets, seeking a veto over the committee's actions, pressed for either unanimous or two-thirds approval for the passage of any measure; the Americans sought to maintain the normal UN committee voting procedure, which required a simple majority for passage. In addition, the Soviets insisted that the chair be a Soviet scientist.

The haggling had three other foci. The Soviets insisted on several proposals which the United States was bound to reject. First, the Soviets wanted to couple space negotiations with talks aimed at general disarmament and the dissolution of foreign military bases.[18] The Americans responded by stating that these were separate issues which they were willing to discuss--separately. Given the facts that any agreement on "complete and total" disarmament would remain a utopian ideal at best for the remainder of the Cold War and that the Soviets were insisting that the Americans simply cede assets which the U.S. held as essential to its national security (Soviet overseas bases were less critical for the land-oriented Soviet sphere of influence; moreover, surely the Eastern European socialist brethren *wanted* the occupying Soviet troops), this was at its essence a stalling tactic undertaken in the hope that it might make the United States seem unwilling to negotiate towards greater world peace.

This prerequisite was also inconsistent with the fact that the Soviets had participated in the International Geophysical Year before any agreements regarding general disarmament had been concluded.[19] This Soviet tactic would resurface (pages 48-49, 54-55 and 69).

Second, the Soviets sought to dictate in law that only government enterprises could conduct space activities. The United States rejected this proposal fully as a restriction on its national sovereignty. This proposal would have restricted only Western countries: while the U.S. space program relied heavily on the private sector, no private space company could arise under Soviet communism.

Third, the COPUOS's disputes centered around military uses of space. The Soviets used the UN Committee as a forum for loudly and repeatedly criticizing the United States for space espionage. In early 1959 Korovin communicated a major shift in Soviet policy on space law, repudiating the applicability of the concept of the "freedom of the seas" to space. Korovin declared that reconnaissance satellites are illegal and claimed that a nation would be justified in taking diplomatic and other "reprisals and retaliation[s]" against them.[20] Georgi Zhukov echoed this policy when, in October 1960 (five months after Gary Powers' U-2 was shot down over the USSR), he warned that "the same considerations of state security" (and thus sovereignty) applied for space and air and that the Soviet Union had the right and ability to do the same to satellites: "If other espionage methods are used, they too will be paralyzed and rebuffed."[21] The Soviets followed with proposals to the UN COPUOS to make space-based reconnaissance illegal on June 7, 1962 and April 16, 1963. Apparently nations no longer revolved under satellites; the new Soviet policy maintained that spacecraft were active, independent objects flying over others' territories, and that, when they were reconnaissance satellites, such overflights were illegal.

Why the contradiction in Soviet policy? The Soviets had declared as recently as 1959 that satellites have free right of passage, as other nations revolve under them. Now they declared that they possessed sovereignty over their air *and* space territory, and would do their best to preserve their sovereignty when spy satellites passed overhead.

In addition, the June 1962 Soviet proposal came less than seven weeks after the USSR launched its first reconnaissance satellite, *Cosmos 4*; the second Soviet proposal came after the Soviet Union had launched a total of six spy satellites.[22] Thus, even after the Soviets began exploiting space for reconnaissance, they insisted that American space espionage was illegal.[23] It was not until 1985 that a Soviet leader (Gorbachev) first publicly acknowledged the Soviet military space program.[24]

This contradiction was a product of self-interest. Although the Soviet space program had rushed ahead to launch larger payloads before the United States, the American space program, from the beginning, heavily emphasized small, durable satellites with advanced abilities--including espionage. The Soviets were undoubtedly concerned about their vulnerability to space observation and had nothing to lose by publicly taking the United States to task for exploiting space in ways they were less capable of. Thanks to their system's closed nature, the Soviets could freely condemn space espionage while the world's public remained ignorant of their expanding space reconnaissance program.

This difference in emphasis was more than a product of superior American technology. It was a product of the different intrinsic needs of the USSR and the United States, and grew out of the priorities the Eisenhower administration implemented in the U.S. space program from its beginning. The Soviet Union, in contrast to the United States, was a closed society. Thus, as previously discussed, the American military depended heavily on reconnaissance. The pressing issue of the time was whether or not the Soviets possessed working ICBMs, and, if so, how many of them there were, where they were and how effective they were. American military planners eagerly awaited spy satellites' results.

Their wait was not a long one. In June 1961, four months after the U.S. successfully launched *Samos 2*, the first successful radio-transmission reconnaissance satellite, the United States officials halved its intelligence estimate of the number of operational Soviet ICBMs to sixty. This news came just on time for the rising tension over Berlin (Kennedy and Khrushchev exchanged sharp words over Berlin in their June 1961 Vienna meeting). By mid-September 1961, having launched five more successful *Discoverer* reconnaissance satellites since the *Samos 2* launch, American officials revised this estimate down to fourteen, and the existence of a missile gap was confirmed--only it was in the opposite direction American leaders had feared![25]

When Khrushchev ordered the construction of the Berlin Wall on August 13, 1961, he demanded that the Western powers sign a peace treaty with East Germany by the year's end. However, the United States, armed with the knowledge of the true nature of the "missile gap," met this demand from a position of power. On October 6, 1961 Kennedy met with Foreign Minister Gromyko, and eleven days later Khrushchev withdrew his ultimatum. Although it is unclear exactly what role the greatly diminished estimation of Soviet strategic strength played in the resolution of this crisis, the U.S.'s persistent espionage efforts (those five *Discoverer* successes were accompanied by seven failures) strongly suggest that America was intent upon gaining valuable intelligence in time for the deepening crisis. Whether or not Kennedy confronted Gromyko with his hard data is not clear, but what is clear is that the United States had taken advantage of the cutting edge of its space technology, and as a result, the United States had gained a powerful bargaining advantage.

"Aggressive" Versus Military Uses of Space

Given this utility, it is no surprise, in retrospect, that the United States' United Nations representatives pushed for the prevention of *aggressive* uses of outer space while the Soviet Union's advocated banning *all military* uses of outer space. American leaders argued that space reconnaissance is not offensive in nature but is instead analogous to ocean surveillance (which both sides accepted as legal).[26] They argued that observation from space would increase world stability by providing a necessary window into the closed Soviet society and laying the foundation for verification of future arms control agreements. After all, the USSR was unlikely to permit on-site verification, and satellite espionage could prove a partial alternative. These positions would be restated in 1963 (page 55).[27]

Meanwhile the Soviet Union worked to catch up in military applications satellites. In 1959 the United States launched five primarily military satellites; it launched ten, nineteen, thirty-one and twenty-six in 1960, 1961, 1962 and 1963, respectively. In contrast, the USSR launched its first military satellites in 1962, when they launched five; in 1963 they launched seven primarily military satellites.[28] Again, the fundamental differences between the two superpowers--one closed and one relatively open--helps explain the continuing difference in policy.

Thus, in one sense the Soviet system of closed government gave the USSR an advantage in the public relations battle. Just as the Soviets could conceal mission failures, just as they would conceal their participation in the "race to the moon" (Chapter Three, page 70), so too could they hide their military space program more easily than the United States could. The fact that the Soviets continued to criticize the American space reconnaissance program after they had their own shows that their complaints were more than stalling tactics designed to allow them to catch up technologically. The Soviet criticisms were a part of the image battle. Even as *Cosmos* satellites photographed American military forces and activities around the globe, the Soviets could, thanks to their outward disapproval of space espionage, demand to know why the Americans, unlike they, were unwilling to demilitarize the heavens and help realize the United Nations goal of avoiding the extension of national rivalries into this new field. The Soviet government could not have made this demand had it not controlled the flow of information within the Eastern Bloc. No such campaign would have ever been possible in a relatively open society like the United States,' where any public crusade berating another nation's space reconnaissance program (had American leaders ever pursued such a campaign) would have surely been met by reporters who, U.S. officials' discussions of the American reconnaissance program in hand, would have cried hypocrisy.[29]

This Soviet secrecy did entail a corresponding image-oriented advantage for the United States: pride in its willingness to show its programs' failures and successes. As Kennedy said after Alan Shepard's May 5, 1961 flight aboard *Freedom 7*,

> I...want to take cognizance of the fact that this flight was made out in the open with all the possibilities of failure, which would have been damaging to our country's prestige. Because great risks were taken in that regard, it

seems to me that we have some right to claim that this open society of ours which risked much, gained much.

One could presume, Kennedy suggested, that the Soviets hid their space failures out of some sort of shame or fear, while the United States bravely overcame difficulties in the public's eye. The American president took more explicit advantage of the contrast between America's "open" society and the Soviet's "closed" in another address later that day, attacking the Soviets for using "the secrecy of the totalitarian state...to hide a wide international network of agents and activities which threaten the fabric of democratic government everywhere in the world."[30] The Soviet space program's secrecy, Kennedy argued, was only a subset of the subversive Soviet system.

A Surge in Attempts at Cooperation

Kennedy had taken office in 1961, and, along with such inflamed anti-totalitarian rhetoric, his administration provided a new burst of activity seeking to prevent the extension of national rivalries into space. In his inauguration address, President Kennedy called for greater superpower cooperation in space, urging "Let both sides seek to invoke the wonders of science instead of its terrors. Together let us explore the stars."[31] In his State of the Union address on January 30, 1961, he stated that "this administration intends to explore promptly all possible areas of cooperation" and that both the U.S. and USSR "would help themselves as well as other nations by removing these [space] endeavors from the bitter and wasteful competition of the Cold War."[32] Cooperation in space was again trying to gain root.

Following Kennedy's proposals, Jerome Wiesner of MIT, Special Assistant to the President for Science and Technology, was assigned to chair a joint NASA-President's Science Advisory Committee-Department of State Panel in February 1961. Although the panel's report advised against "joint projects involving flight of our equipment on Russian vehicles, or vice-versa," it did list twenty-two potential joint ventures, ranging from the simple to the grand, believing that "we should give preference to projects that avoid the difficulties connected with a high degree of involvement or else to projects that are sufficiently bold and dramatic to sweep aside these difficulties." The proposals varied from sharing astronomical facilities to the establishment of an international manned lunar base. Kennedy made an overture regarding these potential cooperative ventures in a telegram to Khrushchev on February 13, but Khrushchev rebuffed the possibilities, insisting that progress in broader arms negotiations would have to precede any cooperation in space.

Khrushchev, undoubtedly confident about the quick progress of the Soviet manned space program (the Soviets were preparing to orbit Gagarin, the first man in space, which they did on April 12, 1961), replied "that favorable conditions for the most speedy solution of these noble tasks...would be created through the settlement of the problem of disarmament."[33]

Meanwhile the COPUOS membership was due to expire on December 31, 1961. The United States took the initiative to break the Committee's logjam. On July 15, 1960 the State Department completed its report, "Position Paper for U.S. Participation in Legal Subcommittee of UN Committee on the Peaceful Uses of Outer Space."[34] This report discussed and drafted a prospective UN Resolution on the peaceful uses of outer space which would advance American interests. Its three major proposals were that the committee be re-invited to study the legal problems associated with the use of space, that celestial bodies be declared exempt from national appropriation and that nations be called upon to publicly register their spacecraft at the UN. On November 27, 1961 the Committee convened a COPUOS meeting at the request of the United States; within a week it presented a draft resolution based largely on this State Department draft.

This draft advanced rapidly, and was unanimously passed by the General Assembly as UN Resolution 1721 (XVI) on December 20, 1961. The resolution includes all three proposals which were contained in the 1960 State Department position paper, and was a major victory for the United States effort. The invitation to consider legal issues had already been issued and was really nothing new, and the declaration that all celestial bodies were unavailable for national appropriation was already de facto true by virtue of practical limitations which extended into the foreseeable future; as such, these two proposals served mostly to advance the multilateral, peaceful image of the U.S.-led initiative. The registration of spacecraft was designed to highlight the relatively open nature of the American program in contrast to the secretive Soviet program.[35] In addition, the resolution formally stated that "international law, including the Charter of the United Nations, applies to outer space and celestial bodies;" as later disagreements would highlight, this statement's meaning was unclear in that international law and the UN Charter have been the subjects of varied interpretations.

The resolution went on to call upon the World Meteorological Organization (WMO) to study how "to develop existing weather forecasting capabilities and to help Member States make effective use of such capabilities" and upon the International Telecommunications Union (ITU) to examine how Members could "make effective use of space communication." These proposals were also spearheaded by American delegates, and focused world attention on areas where the U.S. space program was well ahead of the Soviet Union's.[36] In the years to come the United States would provide current meteorological data and satellite telecommunications to countries around the world, vastly improving its image as *the* country pursuing space for peace.[37]

Finally, Resolution 1721 expanded the Committee's membership from twenty-four to twenty-eight. The United States and Soviet Union each nominated two new members. Moreover, other procedural disagreements were resolved in the resolution's wake. The United States withdrew its insistence that the COPUOS voting procedures reflect the rest of the United Nations,' clearing the way for "consensus" voting (i.e., each Committee action was to be found agreeable by all without necessitating a formal vote). For its part, the Soviet Union withdrew its demand for chairmanship of the Committee.

Shortly after Resolution 1721 was passed NASA published criteria for U.S. international cooperative space activities. These guidelines remained the guiding principles for NASA's cooperative efforts for the ensuing decades:[38]

> NASA's international programs follow guidelines which recognize the interests of U.S. and foreign scientists, establish the basis for sound programs of mutual value, and, contribute substantively and literally to the objectives of international cooperation. The guidelines provide for:
> 1. Designation by each participating government of a central agency for the negotiation and supervision of joint efforts.
> 2. Agreement upon specific projects rather than generalized programs.
> 3. Acceptance of financial responsibility by each participating country for its own contributions to joint projects.
> 4. Projects of mutual scientific interest.
> 5. General publication of scientific results.[39]

The Nature of This Early Cooperation

Resolution 1721 was the benchmark of the first major achievement in U.S.-Soviet space cooperation. The space powers managed to determine a forum (the COPUOS), method (the Committee's members and procedures) and context (international law and the UN Charter) for continued progress and explored specific cooperative measures through super-governmental bodies (the WMO and ITU). Still, this cooperation avoided heavily involved collaboration, constituting what Arnold Frutkin succinctly described as "arm's length cooperation."[40]

This level of cooperation was all that the superpowers were, at the time, comfortable with. The level of cooperation was limited by chilly overall U.S.-Soviet relations. In addition, the mutual aversion to yielding technologies and the shared desire to achieve the image as the most advanced space power most interested in exploiting space for all mankind limited cooperative efforts.

The relative levels of coexisting cooperation and competition were also fueled by the common interests of both nations' space program leaders. Throughout the space age both American and Soviet space program leaders, their own best advocates, maintained and even fanned the flames of competition. This collusion of competition in turn fueled each nation's space budgets. An especially candid acknowledgement of this collusion came when Academician Sedov, Chairman of the Interplanetary Communication Commission of the Soviet Academy of Sciences told NASA Deputy Administrator Hugh Dryden on September 7, 1960

cooperation should be in science programs. If we really cooperated on man-in-space, neither country would have a program because the necessary large support in money and manpower was only because of the competitive element and for political reasons.[41]

As Sedov suggested, cooperation in science programs was easier, partly because it was less expensive and thus did not require budgets fueled by politicians afraid of being outdone by the "reds" or the "imperialists." Cooperation in space science was often easier also because scientists truly sought improved techniques and knowledge through collaboration, and tended to think more in scientific terms and less in political terms. As Vladimir Kurt, Deputy Director of the P. N. Lebedev Physical Institute's Astronomical Space Center described his and his colleagues' frustration, "We are scientists. We would like to cooperate with the Americans, and I think our colleagues in America would like to cooperate with us. If only the politicians would let us...."[42] This sentiment and the relative cheapness fed what were often scientifically productive cooperative ventures in space, including those in space biology (see, for example, pages 79-80 and 98 regarding the publication of the *Principles of Space Biology and Medicine*), astrophysics (as often promoted by the International Astronautical Federation; see pages 72-73, 80 and 98), meteorology (as often promoted by the World Meteorological Organization; see pages 49, 54 and 80), environmental sciences (see, for examples, page 98 on the 1987 U.S.-Soviet agreement regarding cooperation in space, pages 98-99 regarding joint ozone studies, page 100 regarding the Mission to Planet Earth and page 108 regarding additional environmental cooperation) and unmanned missions to planets (see, for example, pages 51-55 regarding the Kennedy-Khrushchev proposals, page 100 regarding mutual tracking of the probe *Galileo* and other deep-space objects and pages 98 and 101-3 regarding the Mars 94/96 missions).

This limited level of cooperation matched what Rand had, in July 1961, recommended the United States adopt as its goals for cooperation in space. As it unfolded, American policy regarding collaborative space efforts closely reflected Rand's advice. This study recommended limited cooperation pursuant to the acquisition of tracking stations, collaboration in data collection, sharing of costs on expensive missions, political gains through the sharing of prestige, evidence of peaceful intentions, creation of an international forum to ascertain the plans of others and spread the truth about America's, presentation of a model for cooperation in other fields, absence of restrictive controls on unilateral activity, aid to developing countries and the opening up of the Communist bloc.[43] Thus, by definition of both countries' limited goals for cooperation in space, cooperation in space had been successful through the end of 1961.

The Kennedy-Khrushchev Exchange

On February 20, 1962, as a part of the Mercury program, the U.S. launched *Friendship 7*, successfully orbiting John Glenn three times in America's first successful full-orbit manned flight.[44] Khrushchev congratulated Kennedy the next day, and, in the congratulatory note, suggested that

> if our countries pooled their efforts--scientific, technical and material--to master the universe, this would be very beneficial for the advance of science and would be joyfully acclaimed by all peoples who would like to see scientific achievements benefit man and not be used for 'cold war' purposes and the arms race.

Kennedy responded on February 21, promising Khrushchev he would respond with specific proposals and, on March 7, after reviewing possibilities with advisors, followed up with a long letter which proposed broad cooperation in space.[45] The President made five specific proposals for cooperation: "the joint establishment of an early operational weather satellite system" with one American and one Soviet satellite, the joint establishment of tracking stations on each others' territories (which were to be manned by the host nations' indigenous technicians), cooperative mapping of the earth's magnetic field (again, with each nation launching one satellite for this purpose), Soviet joining the already-initiated, American-led satellite intercontinental transmissions system and an exchange of space medicine knowledge.

In addition to these specific proposals, Kennedy stated that the United States was "prepared now to discuss broader cooperation in the still more challenging projects which must be undertaken in the exploration of outer space," and proposed that a mutual discussion of long-term space plans could yield further prospective areas for cooperation. Among these potential fields Kennedy specifically mentioned joint manned exploration of the moon and "exhaustive scientific investigation of the planets Mars or Venus, including consideration of the possible utility of manned flights in such programs." Discussions of the countries' long-term plans could begin almost immediately, at the COPUOS meeting scheduled to take place later that month in New York.[46]

Khrushchev wrote back on March 20, declaring that

> all peoples and all mankind are interested in achieving the objective of exploration and peaceful use of outer space...the enormous scale of this task, as well as the enormous difficulties which must be overcome, urgently demand broad unification of the scientific, technical and material capabilities and resources of nations.

Khrushchev told Kennedy that "the direction of your thoughts does not differ in essence from what we conceive to be practical measures in the field of such cooperation," and went on to describe potential cooperation in several areas which Kennedy had mentioned: international satellite communications, weather prediction and the exchange of information on space medicine and the earth's magnetic field. Indeed, meaningful exchanges of space biology would soon take place between the superpowers.

Khrushchev also made two new proposals in his letter. First, he suggested that the United States and Soviet Union cooperate in observing objects launched towards the moon, Mars and Venus. Second, he raised the issue of an "international agreement providing for aid in searching for and rescuing space ships, satellites and capsules that have accidentally fallen."[47]

However, Khrushchev did not respond to Kennedy's specific proposals about launching mutually beneficial satellites to construct systems for predicting weather and studying the earth's magnetic field. Magnetic field studies were particularly important for the USSR, since the magnetic field affected the Soviet Union's ability both to communicate using high frequencies (due to its poor cable system, the USSR used high-frequency communication more than the U.S.) and to jam unwanted foreign radio broadcasts such as Voice of America, Radio Free Europe, Radio Liberty and the British Broadcasting Corporation.[48] In addition, although the Soviet Chairman acknowledged that "international systems of long-distance communication" would provide "a reliable means of communication and hitherto unknown opportunities for broadening contacts between nations," this was a verbal gesture which was never followed up by actual Soviet policy. This is not surprising for two reasons: first, the Soviet Union was a closed society and its leadership depended on control of the flow of information.[49] The Soviet people had a carefully prepackaged image of the West presented to them, and the Soviet government was by no means actually interested in broadening its mass communications contacts with the West.

Second, the communications satellite (comsat) systems which NASA had already begun developing were clearly American-led international communications systems. Eisenhower had approved initial NASA funding for communications satellites (comsats). On May 25, 1961 Kennedy announced that he would accelerate this program, and on July 24, 1961, the administration announced its program. The U.S. would develop one global comsat system as quickly as possible; the government would fund research and development as well as usage of the system in unprofitable, developing countries, launch the satellites, supervise the frequencies, transfer the necessary technologies to developing countries and encourage international participation. Private corporations would compete for participation in a bidding process modeled after Department of Defense contracts. In announcing this policy, Kennedy called for all countries to cooperate "in the interest of world peace and closer brotherhood among peoples throughout the world."[50] Thus, Kennedy, in inviting Khrushchev to cooperate in international satellite communications, was in fact merely specifically addressing this invitation to the Soviet Union--an invitation they could not have wanted to accept. Joining the already begun American-led program would have been tantamount to a recognition of American superiority in comsats.[51]

The Soviets' true colors regarding international comsat cooperation showed forth when the U.S.-led Comsat Corporation sought to obtain a frequency band dedicated to comsats at the October 1963 International Telecommunications Union (ITU) conference in Geneva. The United States wanted the ITU and the International Frequency Registration Board (IFRB) to allocate dedicated frequencies which would encourage and permit governments and private corporations to invest in a global comsat system. Investors, the Americans argued, would not see comsats as commercially viable if there were no guaranteed frequency allocations.

The Soviets responded with a two-pronged plan intended to sidetrack the very satellite communications program Kennedy had invited the USSR to join in his March 7, 1962 letter. First, the Soviets opposed the allocation of specific sections of the spectrum for a nation's exclusive use. Instead they sought to delay partitioning of the spectrum and, second, to empower the ITU, which had a large bloc of Communist members, to regulate all global comsats. Such a surrender of independence was totally unacceptable to the Americans, whose Comsat system was already well developed.[52] As McDougall writes, "Khrushchev showed no interest in a system in which the USSR would have a small minority interest and the United States a monopoly of technology; he denounced the planned consortium as a capitalist tool."[53]

Similarly, the Soviets shunned collaboration in meteorological and magnetic field research. As with comsats, American satellite technology in these two fields was more advanced than the Soviet Union's, and Soviet cooperation would have looked more like capitulation than cooperation among equals. As Khrushchev himself would later admit,

> During my time in office I had influence in the decision not to cooperate in international space research. We were unprepared to do so because we would have had to reveal our methods of operation, our secrets, and our scientific and technical solutions for these problems. We had as yet not reached absolute parity....[54]

Finally, perhaps most foreboding of all in Khrushchev's March 20, 1962 letter was his reminder that

> At the same time it appears obvious to me that the scale of our cooperation in the peaceful conquest of space, as well as the choice of the lines along which such cooperation would seem possible is to a certain extent related to the solution of the disarmament problem. Until an agreement on general and complete disarmament is achieved, both our countries will, nevertheless, be limited in their abilities to cooperate in the field of peaceful use of outer space.

Thus, although Khrushchev did not rule out some "limited" space cooperation, even in this, his height of cooperative enthusiasm, he repeated what was tantamount to a prerequisite to more meaningful cooperation. As aforementioned (Chapter One, pages 13-14), Khrushchev did indeed have reason to hope to disarm. However "general and complete disarmament" would remain a contentious and difficult issue at best for the remainder of the Cold War and, indeed, beyond.

One Step Backward...

Despite Kennedy and Khrushchev's broad proposals for increased cooperation in space and despite their agreement that their UN representatives would negotiate towards the fulfillment of these proposals, actual action pursuant to these goals stalled once again in the Committee on the Peaceful Uses of Outer Space. While the Americans sought specific UN Resolutions, the Soviets sought a broad "Declaration of Basic Principles" and a formal international agreement. The United States proposed one resolution to recommend that states aid and return space personnel and vehicles which land under emergency situations, and a second to create a panel of experts to draft international law regarding space accidents. In contrast, the Soviet Union returned to its demand that space activities be carried out "solely and exclusively by states" (once again barring private "bourgeois" corporations) and proposed that space espionage be explicitly outlawed (the Soviets submitted these proposals on June 7, 1962 and April 16, 1963). The American counter-arguments offered earlier in the Committee's existence (pages 46-47) were repeated, i.e., that

> space observation can contribute to the reduction of the risks of war by accident or miscalculation inherent in dealings with a closed society. And it is a use of space which may prove important some day in monitoring disarmament agreements.[55]

As Leonard Meeker, one of the State Department's delegates to COPUOS, explained the American position in an April 1963 speech,

> International law imposes no restrictions on observation from outside the limits of national jurisdiction.   Observation from outer space, like observations from the high seas or from the air space above the high seas, is consistent with international law....observations from space may in time provide support for arms control....If in fact a nation is not preparing surprise attack, observations from space could help us know this and thereby increase confidence in world security which might otherwise be subject to added and unnecessary doubts.[56]

Why would the United States consider even for a moment yielding the technology that had provided them with invaluable intelligence during the Berlin and Cuban Missile Crises? Why would the U.S. stop doing what it did best while the Soviets continued to launch their own reconnaissance satellites? For their part, why should the Soviets stop their imagery campaign of denouncing the aggressive imperialists if they could still develop their own military spy satellites? After all, by the time the Soviets made their April 1963 proposal, they had already launched five of those very space monsters they publicly sought to outlaw.

Given the common "no reason at all" answer to all three of these questions, not surprisingly, the COPUOS deadlocked. On December 14, 1962 the General Assembly unanimously passed Resolution 1802 (XVII), a resolution remarkable most of all for how little it did. The resolution noted and encouraged the progress concerning the World Meteorological Organization, the International Telecommunications Union and the establishment of a UN-sponsored sounding rocket facility in India.

However, unable to announce any further agreements, the resolution merely noted the several different proposals it had heard and noted "with regret that the Committee on the Peaceful Uses of Outer Space has not yet made recommendations on legal questions connected with the peaceful uses of outer space." Once again the competition to appear as *the* state promoting space for peace had moved to the fore, edging out cooperation.

Notes

1. The Chronology of the Space Age, pages 205-234, shows the specific launches.

It is noteworthy that the American space program's image had the additional burden of having to explain its failures, often in full public limelight, while Soviet failures went unpublicized. Through 1968 the U.S. record improved considerably: it had a total of 196 successful launches and 45 launch failures. *Congress and the Nation* vol. 2 (Washington: Congressional Quarterly Service, 1969) 542.

To the author's knowledge, as of early 1994 there had been no full disclosure by the Russians concerning Soviet space failures, although many scholars, Western and Russian, were attempting to take advantage of the new openness to uncover such unknowns. In a sense this dimmed the luster of *glasnost* (openness), but the reality of the situation in the early 1990s was that the Russian space program had, because of both its role as a source of ever more elusive Russian pride and its increasing reliance on foreign sources of hard currency, a vested interest in perpetuating the shroud that covered the failures of its predecessors. This shroud did start to lift by the beginning of 1996, but some corners have been so carefully concealed or simply lost that the light will never reach them.

2. "Soviet civilian space policy" regards that part of the Soviet space program which was not intended for direct military applications; in fact, the entire Soviet space program was a part of the country's military.

3. Article 36 of the 1944 Convention specifically prohibits the use of "photographic apparatus" in aircraft flying over others' territory; Article 3 states that unauthorized government aircraft may not violate a nation's air space.

4. The Soviets boycotted the 1944 convention because of the participation of Spain, Portugal and Switzerland ("pro-fascist" states which did not recognize the USSR).

5. J. N. Hazard, "The Soviet Union and International Law," *Soviet Studies* vol. 1, no. 3, Jan. 1950: 196.

6. William H. Schauer, *The Politics of Space: A Comparison of the Soviet and American Space Programs* (New York: Holmes and Meier, 1976) 245.

7. G. P. Zadorozhnyi, "*Iskustvennii Sputnik Zemli i Mezhdunarodnoe Pravo*," ("Artificial Satellites and International Law") *Sovetskaia Rossiia* 17 Oct. 1959. See also A. Kislov and S. Krilov (two Soviet legal experts), "State Sovereignty in Air Space," *International Affairs* (Moscow) no. 3, Mar. 1956: 35-44 and E. A. Korovin, "International Status of Cosmic Space," *International Affairs* (Moscow) no. 1, Jan. 1959: 53-59.

8. An October 1950 Rand study considered this question, going so far as to contemplate keeping satellites secret. In the end this possibility was dismissed as being too risky--it was impossible to guarantee secrecy and discovery would prove costly. Kecskemeti, *Problems* 5.

9. On March 15, 1958 the Soviet ambassador to the UN proposed the establishment of a UN agency "to work out an agreed international program for launching intercontinental and space rockets with the aim of studying cosmic space, and [to] supervise the implementation of this program." Arnold W. Frutkin, *International Cooperation in Space* (Prentice Hall: Englewood Cliffs, 1965) 170.

10. The original members of the ad hoc Committee were Argentina, Australia, Belgium, Brazil, Canada, Czechoslovakia, France, India, Iran, Italy, Japan, Mexico, Poland, Sweden, the Union of Soviet Socialist Republics, the United Arab Republic (UAR), the United Kingdom of Great Britain and Northern Ireland and the United States of America. Czechoslovakia and Poland also boycotted the committee; India and the UAR abstained, claiming that nothing could be accomplished without the participation of the two major space powers.

11. McDougall, *Heavens* 256 and Schauer, *Comparison* 251.

12. NSC 5814/1. NHO 27. This document advocated a functional definition of space (i.e., orbiting objects are above the "air" and in space), emphasized the potential for civilian benefits and advocated avoiding a United Nations space agency--just as UN Resolution 1472 did.

13. Dusan J. Djonovich, *United Nations Resolutions* series 1, vol. 8 (Dobbs Ferry: Oceana Publications, 1974) 181.

14. One example of how this concession came to haunt the superpowers occurred in 1972 when United Nations representatives from other nations complained of the exclusive nature of the Soviet and American space programs. The Italian representative asked whether or not "scientists from smaller countries would be allowed to stand in line and buy tickets to board a Soviet or United States manned space laboratory." Kathleen Teltsch, "Space Plans Frustrate the 'Have Nots,'" *New York Times* 14 May 1972: A15.

15. For example, during the same 1972 session, the Swedish and Argentinean representatives sought guarantees against exploitation of a nation's resources by a country which might secretly have satellite information regarding resource location. Teltsch, *Have Nots* A15.

16. Resolution 1472 expanded the Committee's membership to twenty-four members, adding Albania, Austria, Bulgaria, Hungary, Lebanon and Romania. Thus, twelve of the Committee's members were pro-Western, seven were Communist and five were neutral (although the "neutral" United Arab Republic and Lebanon were pro-Soviet at the time).

17. This conference was originally planned for 1960 or 1961 in Resolution 1472 (XIV) "for the exchange of experience in the peaceful uses of outer space." The subsequent disagreements delayed the conference until August 1968. When the conference did take place, "the focus was on space accomplishments to date and on what these could mean to the betterment of mankind....the main thrust was in the direction of highlighting the strictly competitive nature of the U.S. and Soviet space efforts and their results, including efforts related to cooperation with other

nations." Dodd L. Harvey and Linda C. Ciccoritti, *U.S.-Soviet Cooperation in Space* (Miami: Center for Advanced International Studies, 1974) 178-79.

18. This Soviet position of linkage dates back to its original proposal for a UN space committee: on March 15, 1958 the USSR officially recommended to the UN General Assembly that it place on its agenda the "question of the prohibition on using outer space for military purposes, eliminating foreign bases on the territory of other states, and the problem of international cooperation in space research." Edmund Jan Osmanczyk, *The Encyclopedia of the United Nations and International Agreements* (Philadelphia: Taylor and Francis, 1985) 587.

19. Still, Soviet participation in the IGY was not, as is discussed in Chapter Three, pages 73-74, fully candid.

20. Schauer, *Comparison* 251-52.

21. McDougall, *Heavens* 259.

22. The Chronology of the Space Age lists the specific launches. Philip J. Klass, *Secret Sentries in Space* (New York: Random House, 1971) 124-25.

23. Leon Lipson, *Legal Problems of Space Exploration: A Symposium* U.S. Senate, 87th Congress, 1st Session (Document no. 87-26) (Washington: GPO, 1961): 807. G. P. Zhukov, "Problems of Space Law at the Present Stage," *Memorandum of the Soviet Association of International Law at the Brussels Conference of the International Law Association*, Aug. 1962: 30, 35-36. G. P. Zhukov, "Practical Problems of Space Law," *International Affairs* (Moscow) no. 5, May 1963: 28.

24. Smith, *Retrospect* 87. This highly public 1985 admission notwithstanding, Marshal V. D. Sokolovskiy, in *Military Strategy*, described Soviet military space policy in 1962 as follows:

> Soviet military strategy takes into account the need for studying questions on the use of outer space and aerospace vehicles to strengthen the defense of the socialist countries. This must be done to insure the safety of our country, in the interest of all socialist cooperation, for the preservation of peace in the world. It would be a mistake to allow the imperialist camp to achieve superiority in this field. We must oppose the imperialists with more effective means and methods for the use of space for defense purposes. Only in this way can we force them to renounce the use of space for a destructive and devastating war.

Thus, although Marshal Sokolovskiy did not make this statement at a press conference, this document constitutes the admission of the existence of a Soviet military space program by a highly placed military official. Nonetheless, the importance of this admission is diminished by its obscurity (relative to revelations such as the statements by American State Department officials). This document was written five years after Sputnik's launch and was translated into English and published in the West in 1975. Harriet Fast Scott, *Soviet Military Strategy* (New York: Crane and Russak, 1975) 455-58.

A similar dichotomy regarding public discussion of military space programs exists between the highest and other official American representatives. Although some government officials publicly discussed the U.S. space reconnaissance program, no American President openly acknowledged the use of spy satellites until President Carter did so in a July 1979 speech at Cape Kennedy. Major General Robert A. Rosenberg (USAF, ret.), telephone interview, 25 Mar. 1994. General Rosenberg wrote the speech for President Carter.

Even this admission was far from tantamount to "opening up" the secret American satellite programs; for example, the Pentagon did not officially acknowledge the existence of the National Reconnaissance Office (NRO), a highly secret reconnaissance organization, until September 18, 1992. The NRO's 1994 budget was approximately $6 billion--which would be more than 40% of NASA's for the same year. "Secrecy Ends on Spy Agency," *New York Times* 19 Sep. 1992: A5. "Spooked Over Intelligence Cuts," *New York Times* 18 Mar. 1994: A22.

25. See Chapter Three, pages 64-65. Klass 103-7.

26. Don E. Kash, *The Politics of Space Cooperation* (West Lafayette: Purdue University Studies, 1967) 114-18.

27. The utility of space reconnaissance was finally enshrined by both the United States and Soviet Union as a part of the 1972 SALT agreements. Both sides agreed "not to interfere with or use concealment to defeat the purpose of the other side's 'national technical means of verification.'" Herbert F. York, *Arms Control: Readings from Scientific American* (San Francisco: W. H. Freeman, 1973) 260-73. Article XII in ABM treaty and Article V in SALT ("Certain Measures with Respect to the Limitation of Strategic Offensive Arms").

28. Marcia S. Smith, "Space Activities of the United States, C.I.S., and Other Launching Countries/Organizations: 1957-1992," *Congressional Research Service Report for Congress* 93-379 SPR, 31 Mar. 1993: 86.

29. The Gardner and Meeker statements (page 55) are examples of such public discussions.

30. *PP of JFK, 1961* (Washington: GPO, 1962) 366, 368.

31. *PP of JFK, 1961* 2.

32. *PP of JFK, 1961* 2 and 26-27.

33. Harvey and Ciccoritti, *U.S.-Soviet Cooperation* 70-73.

34. Declassified October 15, 1980.

35. In actuality, registration was soon defined to require submission of inclination, apogee, perigee and nodal period data--all of which an amateur astronomer can acquire with a homemade radar set. The symbolic, public relations value of this ostensibly American proposal remained nonetheless. D. S. Greenberg, "Space Notes: Soviet Guests; Restrictions on Military Developments; West Ford," *Science* 20 Apr. 1962: 247.

The U.S. Ambassador to the United Nations also communicated to the UN Secretary General the hope that spacecraft registration might set a precedent for prelaunch, on-site inspections. Adlai Stevenson, "Letter to U. Thant," *Department of State Bulletin* vol. 46, no. 1188, 9 Apr. 1962: 588.

36. Kash, *Politics of Cooperation* 111.

37. These and related activities (e.g., Intelsat work) are discussed in *The Impact of Space Exploration on Mankind*, eds. Carlos Chagas and Vittorio Canuto (The Vatican: Pontificia Academia Scientiarvm, 1986). Relevant articles are: R. R. Colino, "Intelsat: the Global Common Denominator," 99-110. K. D. Hodgkins, "International Cooperation: the Cornerstone of Remote Sensing from Space," 137-56. N. J. Parada, "The Relevance of Remote Sensing to Brazil," 167-86. J. A.

Howard, "Remote Sensing and Food Production for a Growing World Population," 187-206. M. Abdel Hady, "The Use of Remote Sensing and Its Impact on the Economy and Agriculture," 219-36. E. E. Balogun, "Space Science, Weather and Man in Tropical Africa," 237-80. C. Ponnamperuma, "Space Technology and the Conquest of Hunger," 343-64.

38. Arnold Frutkin, personal interview, 28 Jan. 1994. Mr. Frutkin was head of NASA's Office of International Affairs from its inception in 1959 until 1977 and was also a United States representative to the COPUOS technical committee.

39. NASA Office of International Programs, *International Programs* 31 Jan. 1962. NHO 24.

40. Frutkin, *International Cooperation* 100-1.

41. Letter by Hugh Dryden to Administrator Glennan. Milton Eisenhower Library, Johns Hopkins University, Dryden Papers. Glennan was NASA Administrator. McDougall, *Heavens* 350. Such collusion in competition was by no means limited to space science during the Cold War. For a revealing investigation of similar collaboration in competition-fed science budgeting see A. G. Oettinger, "An Essay in Information Retrieval or The Birth of a Myth," *Information and Control* vol. 8, no. 1 (Cambridge: Academic Press, 1965).

42. Personal interview, 10 Aug. 1993.

43. Rand RM-2805-NASA, Jul. 1961, as cited in McDougall, *Heavens* 350-51.

44. This Mercury flight followed suborbital flights of May and July 1961 by Alan Shepard and Virgil Grissom, respectively.

45. Harvey and Ciccoritti, *U.S.-Soviet Cooperation* 86-87.

46. John F. Kennedy, "Text of Letter Dated March 7, 1962 from President Kennedy to Chairman Khrushchev re Cooperation in Peaceful Uses of Outer Space," 7 Mar. 1962. NHO 15.

47. Nikita Sergeevich Khrushchev, "Text of Letter from Chairman Khrushchev in Reply to President Kennedy's Letter of March 7, 1962," 20 Mar. 1962. NHO 16.

48. These broadcasts were Western attempts to break the Soviet government's monopoly of information, i.e., to peel open the closed society. They were extremely popular throughout the Soviet bloc, and as a result they threatened the very base of Soviet power. Official Soviet resentment toward these broadcasts is reflected in the Central Committee's Political Report to the 27th Congress,

> The insidiousness and unscrupulousness of bourgeois propagandists must be countered with a high standard of professionalism on the part of our ideological workers, by the morality and culture of socialists, society, by the openness of our propaganda. We must be on the offensive in exposing ideological subversion and in bringing home truthful information about the actual achievements of socialism, about the socialist way of life.

Dalpat Singh Mehta, *Mass Media in the USSR* (Moscow: Progress Publishers, 1987) 7.

49. For a pithy discussion of the Soviet regime's power base see Mary McAuley, *Soviet Politics, 1917-1991* (Oxford: Oxford UP, 1992). According to McAuley, the Soviet regime maintained its absolute power through four levers: control over "the means of coercion, the attribute of authority, *control of knowledge and ideas*, the

ownership of economic resources" (italics added) 10-11.

50. "Statement by the President on Communications Satellite Policy," 24 Jul. 1961. *PP of JFK, 1961* 530. Congress passed this policy into law on August 31, 1962, officially forming the Comsat Corporation.

51. In fact, the United States was ahead in comsat technology; the U.S. launched *Telstar 1*, AT&T's first satellite, which conducted the world's first live transatlantic TV transmissions on July 10, 1962.

52. Delbert D. Smith, *Communications via Satellite: A Vision in Retrospect* (Boston: A. W. Sijthoff, 1976) 132-35.

53. McDougall, *Heavens* 356-57.   Meanwhile, the Americans continued with their plan to involve international participants; in July 1964 the International Telecommunication Satellite Consortium (INTEL-SAT or Intelsat) was formed by the United States, a consortium of European allies, Canada and Australia. The American Comsat Corporation started with a 61% share in Intelsat; the agreement required that this share remain controlling.

54. Nikita Sergeevich Khrushchev, *Khrushchev Remembers: The Glasnost Tapes* (Boston: Little, Brown and Company, 1990) 188.

55. Richard N. Gardner, "Cooperation in Outer Space," *Foreign Affairs*, vol. 41, no. 2, Jan. 1963: 347. Mr. Gardner was Deputy Assistance Secretary of State at the time he wrote this article.

56. Klass, *Sentries* 125-26.

3

The Context of Competition and Cooperation in Space: Overall U.S.-Soviet Relations and the American and Soviet Space Programs, 1962-1969

We sincerely desire that the establishment of cooperation in the field of peaceful use of outer space facilitate the improvement of relations between our countries, the easing of international tension and the creation of a favorable situation for the peaceful settlement of urgent problems here on our own earth.

> -- Nikita S. Khrushchev in a March 1962 letter to Kennedy[1]

Early Competition and Cooperation, 1957-1962

Thus the first five years of the space age saw the two great space powers fitfully working towards some level of cooperation in the peaceful uses of outer space. They shared a limited amount of data through the International Geophysical Year and agreed to form a United Nations Committee on the Peaceful Uses of Outer Space (COPUOS) only to fail to agree on the Committee's scope and procedures. These disagreements led to a Communist bloc boycott of the Committee. Finally the two sides resolved their procedural differences and passed Resolution 1721 which recognized the applicability of international law and the United Nations Charter to space, invited the COPUOS to further consider space law, declared celestial bodies beyond national appropriation, requested that nations register their spacecraft and called for cooperative efforts with the World Meteorological Association and

International Telecommunications Union. The two superpowers' leaders engaged in a fairly substantive, direct exchange, an exchange which had required specific policy reviews in different sectors of each government; still, the COPUOS was stalled by still further disagreements.

This sequence of events established a pattern of fitful attempts at space cooperation which periodically sputtered over differences that, when considered in the context of two countries competing for world domination, are hardly surprising. In fact, in some ways this world competition had fueled these attempts at cooperation, which, as described, often involved vying to appear to be the superpower most willing to serve its current and potential allies with its activities in space. In striving to better each other, the scientists, engineers and officials of the two nation's space programs gained a mutual respect; this mutual respect helped found a grass roots foundation for later collaboration. Moreover, both Kennedy and Khrushchev had an interest in mitigating the Cold War competition. This, the first era of cooperation in space, laid the foundation for future cooperation.

These coexisting and vying spurts of competition and cooperation are, in combination with these competing motives,[2] well explained by examining broader contexts on two levels. First, there is the framework of overall U.S.-Soviet relations; although time would make it increasingly clear that "complete and general disarmament" need not be a necessary prerequisite to cooperation in space, some level of warmth between the two superpowers was necessary if there was to be any joint work in space. Second, it is important to inspect both nation's overall space programs.

Nuclear Weapons and the Overall U.S.-Soviet Relationship

The space age was, as of the end of 1962, five years old. The Cold War competition took on a whole new meaning when, parallel with the ability to orbit satellites, the superpowers gained the ability to launch nuclear warheads on missiles that could travel anywhere in the world in thirty minutes. Near-instant global reach was married with mass destruction.

Because the technologies to launch nuclear warheads and satellites are so intertwined (for example, the first ICBM and the first satellite were both payloads of the same rocket and the first American satellite was launched on an intermediate-range ballistic missile), it is logical that the two superpowers would be quite guarded in pursuing significant cooperation in space. Neither side wanted to reveal their strengths and weaknesses, and neither side wanted to yield any technologies.

Meanwhile American policy makers tried to determine how substantial the Soviet strategic rocket forces were. Kennedy had struck a chord with the American public when he campaigned on the "missile gap," claiming that the Eisenhower administration had allowed the Soviets to leap ahead in this new, critical military arena. Both the Eisenhower and Kennedy administrations concentrated intelligence efforts on gauging the number and location of Soviet ICBMs often at great risk to those involved and to overall U.S.-Soviet relations (recall, for instance, the downing of the American U-2 reconnaissance plane over the Soviet Union on May 1, 1960 and the subsequent political fallout).[3] And, it bears repeating, the Americans pushed ahead as quickly as possible in developing reconnaissance satellites.

Yet the missile gap had proven to be a falsehood, in fact, an inversion of the truth, by the fall of 1961.[4] Americans had overstated the strength of the Soviet nuclear rocket force. The fact is that, from its birth, Khrushchev's leadership was quite sensitive to the fact that the Soviet Union was in many ways a second-rate superpower. The Soviet leadership was, as the aforementioned bargaining positions evidence, acutely aware that, unlike America, the Soviet Union did not have naval forces and bases spread throughout the world.[5] As a result the USSR was often frustrated in its attempts to exert leverage globally: its military forces were not sufficient to support its political goals. In order to obtain greater leverage the Soviet Union strove towards military parity vis a vis its main adversary, the United States. Still, Khrushchev hoped to cut the overall military budget, and, as such, he turned towards the more economical goal of strategic parity.[6]

For the Soviets achieving some measure of parity involved two steps: first, the actual achievement of parity; second, American recognition of this achievement.[7] This two-step process augments an understanding of why the Soviets were so resistant to joining American-led cooperative space ventures and why, on the other side, they were so eager to assume the position of and be recognized as the world's space leader (recall, for example, their original insistence on the chairmanship of the COPUOS).

Khrushchev knew very well that he was overstating the Soviet strategic strength. His foreign policy was marked by risky bluffs followed by retreats, such as those involved with the Berlin and Cuban Crises. Consider, for example, Khrushchev's later comments on his boast about Soviet anti-ballistic missile weaponry: "I made speeches to bolster the morale of my people. I wanted to give our enemy pause by saying that we had antimissile weaponry. I exaggerated a little. I said that we had the capability of shooting a fly out of space with our missiles."[8]

Trouble in Cuba

Cuba was a critical Soviet ally, its closest in the Western hemisphere, and was strategically valuable, located only ninety miles from the Florida coast. In 1961 the United States had stepped up its anti-Castro campaign, exerting overt political and economic pressures, leading the disastrous Bay of Pigs invasion and undertaking clandestine subversive operations against the country, including assassination plots again Castro.

It was in this already highly charged context that Khrushchev, in the spring of 1962, decided to place medium-range nuclear missiles in Cuba. Nuclear missiles in Cuba would go a long way--quickly and relatively cheaply--towards achieving strategic parity. "Khrushchev stressed the need to find 'effective means of deterrence' because he was sure the Americans would invade [Cuba], but also favored the missile deployment 'to repay the Americans in kind' for surrounding the Soviet Union with military bases and missiles, 'to learn what it's like to live under the sights of nuclear weapons.'"[9]

On October 16, 1962 Kennedy was shown U-2 pictures of Soviet missile and airport sites under construction.[10] Six days later the president dramatically announced the existence of the Soviet missile construction sites and announced a naval blockade. For the thirteen days between the missiles' discovery and the crisis' resolution the two superpowers stood at the brink of nuclear war which threatened the lives of literally hundreds of millions of people. Finally, on October 28 the Soviets announced that they would withdraw the missiles from Cuba, and a direct superpower engagement was averted.

Such was the context of the sputtering of the Committee on the Peaceful Uses of Outer Space towards the lackluster UN Resolution 1802. Given that the two superpowers stood on the brink of all-out war, it is hardly surprising that they could not agree on space law and other space cooperation issues. Similarly, technical discussions aimed at establishing cooperation sputtered. Deputy Administrator Dryden and Soviet Academician A. A. Blagonravov found their discussions strained at best and were forced to explore only modest proposals.[11]

The Moon Race

Meanwhile the American space program was heading into its single largest period of expansion. During the 1960 presidential campaign Kennedy asserted that,

We are in a strategic space race with the Russians, and we have been losing....Control of space will be decided in the next decade. If the Soviets control space they can control earth, as in past centuries the nation that controlled the seas dominated the continents....we cannot run second in this vital race. To insure peace and freedom, we must be first.[12]

Six months after this article appeared the first man orbited the earth--and that man was a Soviet cosmonaut. When Yuri Gagarin was the first man to orbit the earth on April 12, 1961, Sputnik-like echoes made themselves heard across the world--especially in the United States. Once again the Soviet Union had beaten the United States in an impressive space feat. Consider, as an example of the U.S. reaction, Representative Fulton's (R - Pennsylvania) remarks at a special hearing before the House Space Committee:

I believe we are in a race, and I have said many times, Mr. Webb, 'Tell us how much money you need and we on this committee will authorize all you need....' I am tired of being second to the Soviet Union. I want to be first....Do you gentlemen realize you have the responsibility of the way the capitalistic system looks to the world--on its efficiency and on its progress in scientific development?...Don't you think the Soviet Union having the first man in space gives them a tremendous advantage at the bargaining table...[that] this has placed immense strength...in Soviet hands?

In response NASA Administrator Webb and Deputy Administrator Dryden assured the Committee that the Mercury program was working on an "urgent basis," "moving as rapidly as we know how to move" with double and triple shifts.[13] The limiting factor, the American leaders made sure, was not money; it was time that was of the essence. If there really was a "vital race" for "control of space," as the President had stated, it seemed that the United States was running second.

For Kennedy, the timing of Gagarin's orbit could hardly have been worse. The Bay of Pigs disaster came only three days later, on April 15, 1961. The United States looked weak in the world's eyes; Kennedy seemed to be the all-too young mastermind of this weakness. According to Theodore Sorenson, Kennedy was influenced by

the fact that the Soviets had gained tremendous world-wide prestige from the Gagarin flight at the same time we had suffered a loss of prestige from the Bay of Pigs. It pointed up the fact that the prestige was a real, and not simply a public relations, factor in world affairs.[14]

Kennedy had repeatedly asserted during the presidential campaign that the United States and the Soviet Union were locked in a deadly race for world domination. The United States, Kennedy insisted, must accept this fact and take the lead. After Gagarin's flight America's image continued to languish in the trail of Soviet space achievements, and Kennedy's promises seemed empty. Reacting to the public relations debacle quickly, Kennedy impatiently asked in an April 14 meeting, "Is there any place we can catch them? What can we do? Can we leapfrog?...When we know more, I can decide if it's worth it or not. If somebody can just tell me how to catch up...."[15]

In order to examine the path needed for such a leapfrog, the President ordered Vice President Johnson to reevaluate America's space efforts, asking if

we have a chance of beating the Soviets by putting a laboratory in space, or by a trip around the moon, or by a rocket to land on the moon, or by a rocket to go to the moon and back with a man. *Is there any other space program which promises dramatic results in which we could win?*...How much additional would it cost....Are we working twenty-four hours a day on existing programs. If not, why not?...Are we making maximum effort?[16]

In this same memorandum the President stated that he had ordered NASA Administrator Webb and the Secretary of Defense, Robert McNamara, to cooperate with the National Aeronautics and Space Council (as chaired by Lyndon Johnson). Kennedy made it clear that something must be done, and all indications were that it would be something dramatic.

On May 8 Webb and McNamara responded with a report to the Vice President. Webb wrote

the dramatic achievements in space which occurred in recent weeks should have made clear to us all the impact of this new frontier of human adventure....Now it is time to take longer strides--time for a great new American enterprise--time for this nation to take a clearly leading role in space achievement...

At this point the Vice President added by pen

...which in many ways may hold the key to our future on earth.[17]

The report included specific NASA and Department of Defense proposals, and Johnson forwarded them to Kennedy that very day.

Such was the heated context of Kennedy's decision to initiate the Apollo manned lunar landings. Thus the United States entered and raised the stakes of the space race in earnest when, on May 25, 1961 Kennedy declared in his message on "Urgent National Needs,"

Finally, if we are to win the battle that is going on around the world between freedom and tyranny, if we are to win the battle for men's minds, the dramatic achievements in space which occurred in recent weeks should have made clear to us all, as did the sputnik in 1957, the impact of this adventure on the minds of men everywhere who are attempting to make a determination of which road they should take....Recognizing the head start obtained by the Soviets....we nevertheless are required to make new efforts on our own....I believe that this Nation should commit itself to achieving the goal, before this decade is out, of landing a man on the moon and returning him safely to earth. No single space project in this period will be more exciting, or more impressive to mankind...and none will be so difficult or expensive to accomplish.[18]

Any hopes for cooperation took a further blow during the Kennedy/Khrushchev summit in Vienna. Although the summit lacked a formal agenda, Kennedy asked Khrushchev point blank if he thought the United States and Soviet Union should go to the moon together. Khrushchev bluntly replied that cooperation in space was impossible without disarmament.[19] Another Berlin Crisis (and the building of the Berlin Wall on August 13, 1961) and a Soviet test of a nuclear weapon (breaking the moratorium on September 1, 1961) soured relations further. Efforts in space were moving forward quickly, but by no means together.

That year began a skyrocketing of the U.S. space budget. NASA's fiscal year 1962 budget of $1.1 billion was increased by $665 million. Similarly, for the fiscal year 1962 the Department of Defense received $850 million dollars for space, a $226 million increase over the previous year.[20] NASA built the Manned Spaceflight Center near Houston and facilities at Cape Canaveral in Florida (including the world's largest enclosed building, the Vehicle Assembly Building), the Mississippi Test Facility, a research and development center in Michoud, Louisiana and expanded operations across the country. By 1964 the NASA budget burgeoned to $5.1 billion. Kennedy was determined that the United States would make the maximum effort to be the world's unquestioned leader in space, and the question about resources was not "If only we had...?" but "How much do we need?"

Between 1962 and 1969 the United States developed the *Gemini* and *Apollo* spacecraft and achieved several "firsts" in space. During this period the U.S. launched the first maneuverable spacecraft, operational geostationary comsat, orbiting astrophysics observatory, orbiting solar observatory, internationally cooperative satellite and micrometeorite satellite. American *Ranger*, *Orbiter* and *Surveyor* spacecraft mapped the entire lunar surface. The U.S. developed a network of comsats over the Atlantic and Pacific oceans with three generations of *Intelsat* satellites, set a record of 206 manned orbits with *Gemini 7*, developed the massive Saturn 5 rocket (which orbited a record 278,699 pounds) and conducted the first ship-to-ship spacewalk and the first successful close-up photographing of Venus and Mars. America began the world's first satellite weather system and completed the first operational satellite navigation system during this era. Finally, on July 16, 1969, the U.S. launched *Apollo 11*, and on July 20, 1969, Neil Armstrong became the first man to step on the moon.

The Soviets also vastly increased their space efforts. Between 1962 and 1969 they developed the *Soyuz* spacecraft, launched two satellites within twenty-four hours of each other, launched their first reconnaissance, communications and weather satellites, conducted unmanned missions to Mars and Venus, set a manned longevity record, launched the first woman and the first doctor into space, conducted the first space walk and orbited a record three manned spacecraft at the same time. The USSR also began developing the first co-orbital antisatellite (ASAT) device and founded Interkosmos, a part of the Academy of Sciences dedicated to cooperative space ventures with other socialist nations. Korolev, the father of the Soviet space program, died in 1966.

The Soviets also secretly joined the "moon race." The Soviet manned lunar program faced severe setbacks when two of its unmanned *Zond* lunar missions failed in February and March 1967, and especially when the booster being built for the lunar mission exploded in June 1969 (there were apparently earlier failures of this booster; it would fail again in June 1971 and November 1972).[21] After the Apollo landings the Soviets stated categorically that they were never engaged in any moon race, and that they were instead accomplishing more science at less expense and danger with their unmanned lunar program.

However, in 1989 Soviet officials disclosed that they *had* been trying to beat the United States to the moon. On December 11, 1993, the Soviet space program, pressed by extreme financial need, auctioned the 31-volume diaries of the former head of the Soviet space program, Dr. Vasily P. Mishin; the diaries covered the period from 1957 to 1974 and confirmed the Soviet lunar attempts. Also sold at the Sotheby's auction was a Soviet spacesuit intended for a moon landing.[22] It is now publicly known that the Soviets were unable to successfully develop a booster powerful enough for a piloted lunar mission (it is unclear what Kennedy, Johnson and other U.S. officials may have secretly known about the Soviet efforts).

The *Apollo 11* mission and the subsequent five manned lunar landings were a major public relations victory for the American space program.

Close followers of the leading countries' space programs would have seen the successful July 1969 moon landing as particularly poignant; even as *Apollo 11*'s astronauts orbited the moon, preparing for touchdown, the Soviet *Luna 15* crashed on the lunar surface, failing in an attempted soft landing.

Together to the Moon?

An improvement in the overall international climate which started in mid-1963 allowed even these, the years of the moon race, to be marked by American consideration of greater cooperation and by significant progress in the formation of space law. On August 5, 1963, the United States and Soviet Union signed the Treaty Banning Nuclear Weapon Tests in the Atmosphere, Outer Space and Under Water.[23] Meanwhile both America's and the Soviet Union's relations with China were worsening considerably: China was becoming more involved in the increasingly intense Vietnam conflict, and was at the same time becoming more assertive in what would become the open Sino-Soviet split. As a result, the U.S. and USSR each sought to ease their relations in order to increase their security vis a vis China. Soviet-American tensions had lessened considerably since the Gagarin flight, Berlin Crisis, breaking of the nuclear testing moratorium and Cuban Missile Crisis. Still, the arms race not only continued but seemed to quicken; both countries stood to gain from an attenuation of this costly competition.

At the same time the U.S. manned moon program, Project Apollo, came under attack as too expensive and unnecessary. This pressure, combined with the changed international climate, allowed for more realistic forays into the kinds of cooperation Kennedy had suggested ought to be possible during his election campaign. Kennedy hoped to mitigate the cold war tension by initiating U.S.-Soviet cooperation: "Wherever we can find an area where Soviet and American interests permit effective cooperation, that area should be isolated and developed....certain aspects of the exploration of space might be handled by joint efforts."[24]

In pursuit of this goal, on September 18, 1963, President Kennedy met with Administrator Webb to discuss the possibility of turning the Apollo program into a cooperative venture with the Soviet Union. Although this possibility had been studied earlier, this step by the Chief Executive marked a dramatic shift from the program's original motivation, i.e., as a blatant "Let's beat the Soviets as dashingly as we can!"[25] In a memorandum regarding that meeting National Security Advisor McGeorge Bundy told Kennedy, "If we cooperate, the pressure comes off, and we can easily argue that it was our crash effort in '61 and '62 which made the Soviets ready to cooperate. I am for cooperation if it is possible, and I think we need to make a really major effort inside and outside the government to find out whether in fact it can be done." Bundy concluded that even this, the biggest unilateral space venture ever, should be turned into a superpower cooperative project if it proved feasible.[26]

Two days after his meeting with Webb, Kennedy called for a joint U.S.-Soviet manned mission to the moon in a speech delivered to the United Nations General Assembly:

> Why, therefore, should man's first flight to the moon be a matter of national competition? Why should the U.S. and Soviet Union, in preparing for such expeditions, become involved in immense duplications of research, construction, and expenditure? Surely we should explore whether the scientists and astronauts of the two countries--indeed of all the world--cannot work together in the conquest of space, sending some day in this decade to the moon not the representatives of a single nation, but the representative of all our countries.[27]

On November 12 in National Security Action Memorandum Number 271 the President asked Webb to develop proposals "with a view to their possible discussion with the Soviet Union as a direct outcome of my September 20 proposal for broader cooperation between the United States and the USSR in outer space, including cooperation in lunar landing programs."[28] Ten days later, on November 22, 1963, Kennedy was assassinated.

On January 21, 1964, NASA's Deputy Administrator, Hugh Dryden, suggested to President Johnson specific, smaller-scale cooperative programs which might build up to a joint manned lunar landing. Adding a note of realism, he advised Johnson that "adequate implementation" of even modest space cooperation "requires a change in the Soviet attitude with respect to disclosure of information with regard to program plans and results....this [limited, step-by-step] proposal constitutes a practical step which goes as far as we should prudently go in the light of our own national interests."[29]

Not surprisingly, the Soviets did not respond favorably to these overtures. They were quite busy with their own two-fold plan: publicly denying any interest in wasting the money to irresponsibly risk landing men on the moon while secretly racing to beat the Americans there. The American program was begun with great fanfare and, as Bundy pointed out, by September 1963 the Apollo Program was in fact thoroughly under way. There was simply no way the Soviets could have cooperated without being described as joining the effort as secondary partners. Moreover, the Soviets' caution derived partly from their focus on internal affairs: on October 14, 1964, the Central Committee deposed Khrushchev, and Brezhnev assumed Party leadership.

Progress in the UN

Greater Data Exchanges and the Treaty on the Use of Outer Space

Although no U.S.-Soviet manned moon mission was in the stars, the improved bilateral climate did facilitate greater agreement in the United Nations. This greater concord took two forms: greater exchanges of scientific and data and progress in space law.

The first area of increased cooperation was in the exchange of scientific data, and was rooted in American frustrations with the Soviets' reluctance to share information during the International Geophysical Year (July 1, 1957 to December 31, 1958). As Arnold Frutkin, who was the Deputy Director of the U.S. National Committee for the International Geophysical Year before he worked at NASA, recalled, American scientists and delegates frequently confronted their Soviet counterparts during the IGY, asking where their promised data was:

'Where's your stuff? This was supposed to be an open program. You announced an open program. This is the *International* Geophysical Year.' And of course you got really nothing from them.

During 1963 the American delegates to the technical subcommittee of the COPUOS changed their tactics, successfully securing greater Soviet scientific cooperation. As Frutkin explained,

We finally did embarrass them into cooperation. There was a specific device used....I thought that...it doesn't get us anywhere to say we're cooperating and you're not. You need a sort of dossier. So we began writing them letters deliberately, proposing cooperation. Many of those letters were never answered. Some were answered with the most ridiculous fobbing off. Until we had a dossier. We had a long list..one or two type-written pages. We then proceeded to use it publicly...at any meeting we could. If I had a talk to give at say, an International Astronautical Federation meeting...I would hold up this list and go through it, saying on such-and-such a date we wrote, no answer. On such-and-such a date, we wrote proposing so-and-so, answer: 'No.' We embarrassed the hell out of them. Some of the people there representing them would come to me after the meetings and say, 'Mr. Frutkin, you're being very, very unfair.' I would say, 'Is there anything here which isn't truthful?' 'No, but we can't help that, though.' 'Well, it's the truth.' Very soon after that we got a letter and we started cooperating.[30]

This development is an excellent example of one of the disadvantages of the Soviet closed system. When promoting itself in a multilateral forum as a promoter of the exploitation of space for all mankind, the inbred Soviet resistance to openness became an all-too noticeable liability. Candor and forthrightness have an intrinsic advantage over reticence and restraint in an open, international arena characterized by and working towards trans-national exchanges. The chipping away for information from the Soviets had begun, facilitated by an improvement in relations and wrought through a systematic campaign of embarrassment. This process would continue to be a slow, arduous one, one which would face setbacks in the late 1970s/first half of the 1980s, and would even outlast the Soviet Union itself.

The second area of greater cooperation begun in 1963 was progress in the establishment of international space law. The foundation for this progress was set when on October 17, 1963 UN Resolution 1884 (XVIII) passed unanimously, banning nuclear weapons from space.[31]

On December 13, 1963 UN Resolutions 1962 and 1963 (XVIII), the "Declaration of Legal Principles Governing the Activities of States in the Exploration and Use of Outer Space" and "International Cooperation in the Peaceful Uses of Outer Space," respectively, were adopted unanimously. The resolutions established the beginnings of space law. Resolution 1962 proposed the following:

1. That "the exploration and use of outer space...be carried on for the benefit...of all mankind."
2. That all states may freely explore space.
3. That "space and celestial bodies are not subject to national appropriation."
4. That international law and the UN Charter apply to outer space.
5. That all states shall assist and return, in the event of an emergency, astronauts and their vehicles, treating all "astronauts as envoys of mankind."
6. That states have responsibility for any (i.e., private or governmental) national space activities.
7. That states shall be liable for any damage associated with their use of space.
8. That states shall retain "jurisdiction and control" over any object they launch.

Resolution 1963 called for the COPUOS to continued to work towards a more formal "international agreement" while outlining several new provisions:

1. An obligation to inform the public, to the extent feasible, of the nature, conduct, location and results of space activities ("to the extent feasible" was understood to allow individual nations to report data as they saw fit, i.e., voluntarily).
2. A prohibition against the orbiting of weapons of mass destruction or the use of celestial bodies for military bases.
3. An obligation to consider on the basis of equality any request to use tracking facilities or provide land for tracking facilities.
4. An obligation that all installations on celestial bodies shall be open on the basis of reciprocity to other states.
5. Acknowledgement of the fact that space activities are the responsibility jointly of member states and international intergovernmental organizations when the activities are carried on by such states/organizations.

In negotiating this treaty, both the Americans and the Soviets made concessions. The United States accepted the Soviet position that data reporting be voluntary (the American proposal had made this mandatory) and agreed to the Soviet-proposed inclusion of the clause preventing the placement of weapons of mass destruction into space (this clause had already appeared in UN Resolution 1884). In addition, the Americans dropped their proposal that all space stations and vehicles be freely available for inspection. For their part, the Soviets dropped their insistence that all states have equal access to others' tracking facilities; this clause was highly impractical (the host states would have undoubtedly resisted this violation of their sovereign right to determine who could utilize their territory) and would have given the Soviets access to the hard-won American tracking network (this network included American stations in more than twenty countries, and all stations were arranged through bilateral agreements).[32] Lastly, the clause holding governments responsible for their country's private space activities represented a compromise by both sides: the Soviet Union acknowledged non-governmental entities' right to pursue space endeavors while the United States assumed liability for such activities.

The next step was the elevation of these agreed-upon principles to a formal "Space Law," i.e., to treaty level. This step proved to be not without its difficulties. Responding to the lack of progress between December 1963 and December 1965, the General Assembly requested that the logjammed COPUOS "continue with determination the preparation of draft international agreements...[and] give consideration to incorporating in international agreement form...appropriate legal principles governing the activities of States in the exploration and use of outer space."[33] President Johnson gave the process new impetus when he called for a quicker agreement on space law, stating that outer space."[33] President Johnson gave the process new impetus when he called for a quicker agreement on space law, stating that

> Just as the United States is striving to help achieve peace on earth, we want to do what we can to insure that explorations of the moon and other celestial bodies will be for peaceful purposes only....We want the results of these activities to be available for all mankind....I am convinced that we should do what we can--not only for our generation, but for future generations--to see to it that serious political conflicts do not arise as a result of space activities. I believe the time is ripe for action. We should not lose time. I am asking Ambassador Goldberg [U.S. representative to the United Nations], in New York, to seek early discussions of such a treaty in the appropriate United Nations body.[34]

Following this executive push the United States and Soviet Union proposed draft resolutions to the United Nations one after the other in May 1966. Their differences were resolved and finally, on December 19, 1966, the General Assembly passed UN Resolution 2222 (XXI). The resolution included the draft "Treaty on Principles Governing the Activities of States in the Exploration and Use of Outer Space, Including the Moon and Other Celestial Bodies."

This draft was formally signed as an international treaty in Moscow on January 27, 1967. The resulting "Treaty on Principles Governing the Activities of States in the Exploration and Use of Outer Space, Including the Moon and Other Celestial Bodies" entered into force on October 10, 1967.[35] Although the Treaty included nothing new, it raised all of the provisions of Resolutions 1962 and 1963 to treaty level, and marked the end of the first era of attempts to arrive at some consensus on international space law.

Notes

1. Nikita Sergeevich Khrushchev, "Text of Letter from Chairman Khrushchev in Reply to President Kennedy's Letter of March 7, 1962," 20 Mar. 1962: 4. NHO 16.

2. The events summarized here are discussed in greater detail in Chapter Two.

3. Chapter Two, pages 44-45.

4. Chapter Two, pages 45-46.

5. Chapter Two, page 44.

6. Chapter One, pages 12-14.

7. Raymond L. Garthoff, *Détente and Confrontation: American-Soviet Relations from Nixon to Reagan* (Washington: The Brookings Institution, 1985) 59-67.

8. Khrushchev, *Glasnost Tapes* 188.

9. Raymond L. Garthoff, *Reflections on the Cuban Missile Crisis* (Washington: The Brookings Institution, 1989) 15. Garthoff is quoting Alekseyev, *Ekho Planety* Nov. 1988: 28-29.

10. As far as is known, the United States did not employ satellite reconnaissance over Cuba as a part of the Cuban Missile Crisis. U-2 flights easily performed satisfactory espionage from bases in Florida, while satellites were concentrated on photographing the Soviet Union. Klass 121.

11. Edward Clinton Ezell and Linda Neuman Ezell, *The Partnership, A History of the Apollo-Soyuz Test Project* (Washington: National Aeronautics and Space Administration, 1978) 45-46.

12. John F. Kennedy, "If the Soviets Control Space, They Can Control the Earth," *Missiles and Rockets* 10 Oct. 1960: 12.

13. U.S. House, Committee on Science and Astronautics, *Discussion of Soviet Man in Space Shot* 87th Congress, 1st Session, 13 Apr. 1961 (Washington: GPO, 1961) 7-9.

14. John M. Logsdon, *The Decision to Go to the Moon* (Chicago: The University of Chicago Press, 1970) 112. From an interview by Logsdon of Sorenson.

15. Harvey and Ciccoritti, *U.S.-Soviet Cooperation* 75.

16. John F. Kennedy, "Memorandum for the Vice President," 20 Apr. 1961. NHO 14.

17. James E. Webb and Robert McNamara, "Recommendations for Our National Space Program: Changes, Policies, Goals," 8 May 1961. NHO 35.

18. *PP of JFK, 1961* 403-4.

19. Harvey and Ciccoritti, *U.S.-Soviet Cooperation* 78-79.

20. McDougall, *Heavens* 335.

21. Oberg, *Star* 117-22.

22. John Noble Wilford, "Soviet Space Papers Going on Sale," *New York Times* 5 Dec. 1993: A36. The Russians also sold a *Soyuz* spacecraft from a 1990 mission for $1.65 million. "What's Up? Ask an Auctioneer," *New York Times* 24 Feb. 1994: C1. "The Russians May Have Succeeded in Sheltering Artifacts," *Space News* 3-9 Jan. 1994: 4.

23. This treaty is also known as the "Limited Test Ban" or "Partial Test Ban" Treaty, and was signed in Moscow in the presence of the United Nations Secretary General.

24. John F. Kennedy, *Bulletin of the Atomic Scientists* 347.

25. The American idea of proposing a joint piloted moon mission to the Soviets was first discussed in March 1961 by the joint NASA-President's Science Advisory Committee-Department of State Panel. The panel recommended offering the Soviet Union a joint manned mission to the moon, arguing that if they refused, "the United States would be no worse off, from a propaganda point of view, for having made this offer at this time." Harvey and Ciccoritti, *U.S.-Soviet Cooperation* 68.

26. McGeorge Bundy, "Memorandum for the President," 13 Jul. 1962. NHO 2.

27. *PP of JFK, 1963* (Washington: GPO, 1964) 695.

28. John F. Kennedy, "Memorandum for The Administrator, NASA," 12 Nov. 1962. NHO 13.

29. Hugh L. Dryden, letter to the President.

30. Arnold Frutkin, personal interview, 28 Jan. 1994.

31. UN Resolution 1884 (XVIII), "Question of General and Complete Disarmament," called upon all states to "refrain from placing in orbit around the earth any objects carrying nuclear weapons or any other kinds of weapons of mass destruction. Djonovich, vol. 9: 203.

32. Kash, *Politics of Cooperation* 120-21.

33. UN Resolution 2130 (XVIII), "International Cooperation in the Peaceful Uses of Outer Space," December 21, 1965. Djonovich, *Resolutions* vol. 10: 106-7.

34. *Public Papers of the Presidents of the United States: Lyndon B. Johnson, 1966* (Washington: GPO, 1967) 487-88. May 7, 1966 speech in San Antonio, Texas.

35. United Nations, *Treaty Series* vol. 610 (New York: United Nations, 1971).

4

Cooperation Come and Gone: The 1975 Apollo-Soyuz Test Project and the Subsequent Souring of Superpower Relations, 1969-1980

Now, what is the relation between such joint enterprises and the reduction of danger of war? If they work, they help change attitudes. Men who work side by side effectively learn that the human race has common traits and interests which transcend national boundaries and rivalries.

-- John F. Kennedy in a November 1960 public letter[1]

The Ice Is Cracked

The second era of cooperation in space began formally on May 24, 1972 when President Nixon and Premier Kosygin signed the "Agreement Concerning Cooperation in the Exploration and Use of Outer Space for Peaceful Purposes." In this landmark agreement, the two space powers agreed to "develop cooperation in the fields of space meteorology, study of the natural environment, exploration of near earth space, the moon and the planets, and space biology and medicine."[2] In addition, both countries committed to encouraging and facilitating "the establishment and development of direct contacts and cooperation between agencies, organizations and firms."[3] As Soviet Academician Boris N. Petrov, then Chairman of USSR Interkosmos Council, described the accord,

it is difficult to overestimate the significance of the agreement which has been concluded....outer space is becoming, in all its aspects, an arena for broad international cooperation and demands joint efforts of many countries, especially those countries which already have made considerable achievements in this matter.[4]

The 1972 agreement facilitated unprecedented scientific exchanges. As a result of the agreement on space biology, U.S. and Soviet experts (led by Soviet Academy of Sciences Oleg Gazenko and American Nobel Prize Winner Melvin Calvin) collaborated in the July 1975 publication of a three-volume *Principles of Space Biology and Medicine*. This allowed the scientists of the two leading space programs to exchange information they had acquired, including that gleaned from the space station experiences of Skylab and Salyut's 1 and 3. In addition, U.S. and Soviet scientists exchanged papers and information at a colloquium on solar system chemistry from June 4-8, 1974 in Moscow. There scientists exchanged data and relevant theories based on observations from terrestrial facilities, the American Apollo and Mariner missions and the Soviet Luna and Venera probes. Topics covered included comparisons of lunar and planetary materials, the thermal history of the moon, lunar gravitation and magnetism, the evolution of the moon, planets and meteorites, the formation of the lunar surface, chemical theories regarding the origin of the moon and planets and new data about Mercury, Venus, Mars and Jupiter. These were astronomers and geophysicists accustomed to squeezing as much science as they could from limited amounts of data, and the exchange provided a novel opportunity for the world's two leading planetary science programs, heretofore working in near isolation from one another, to compare information, methods and conclusions. As Noel W. Hinners, NASA's Associate Administrator for Space Science, judged, "significant exchanges of ideas and data did occur on the personal level" during the conference, due largely to "a clear recognition by all concerned of the scientific contributions by both the United States and the USSR and a gratifying lack of protocol and paperwork."[5]

Most importantly, this agreement committed the United States and the Soviet Union to a joint docking between the Apollo and Soyuz spacecraft. Thus was born the Apollo-Soyuz Test Project (ASTP).[6] For the first time American and Soviet astronauts, cosmonauts and hardware would meet in space. Joint training exercises and technical exchanges of an unprecedented level would be required; by the time the mission took place, there had been forty-four working group meetings and joint communications and tracking exercises.[7] Perhaps the most important aspect of the preparations was that it involved the first large-scale exchanges of Soviet and American space scientists, engineers and officials: approximately one hundred officials from each nation visited the others' country, establishing contacts, exploring the different environments and exchanging ideas on an unprecedented level.[8]

The specific plans for a U.S.-Soviet docking stood out in stark contrast with the image competition, limited exchanges and exhaustive haggling (especially over space law) of the previous fifteen years of the space age. As The *Wall Street Journal* stated, "[a]fter years of competition and years of almost-fruitless talks about cooperation, both sides are moving toward collaboration on specific projects."[9]

The Improved International Climate

What made this agreement possible? How was it that now the superpowers had finally concluded a substantive program of space cooperation? The difference was détente.

U.S.-Soviet relations in the early 1970s were characterized by détente. The May 1972 agreement was signed in the wake of agreements on space law[10] and the Partial Test Ban, Outer Space, Nuclear Non-proliferation and Strategic Arms Limitation (SALT-1) Treaties. In addition, an increasingly assertive China again pushed the two superpowers closer together. The Sino-Soviet rift had opened up into armed border conflicts in 1968 and 1969; the Soviet Union sought some rapprochement with the United States to prevent China from becoming too close with the western superpower. Nixon and Kissinger did their best to legitimize this Soviet fear, travelling to China just before heading to Moscow for the 1972 summit.

Spectacular Cooperation

After three years of preparation the American Apollo and Soviet Soyuz spacecraft approached each other on July 17, 1975. At 8:05 a.m. (Houston time) the two craft, carrying astronauts Stafford, Brand and Slayton and cosmonauts Leonov and Kubasov, respectively, established radio contact. The following simple conversation (held in English) began the landmark link-up:

Slayton: Soyuz, Apollo. How do you read me?
Kubasov: Very well. Hello everybody.
Slayton: Hello, Valeriy. How are you. Good day, Valeriy.
Kubasov: How are you? Good day.
Slayton: Excellent....I'm very happy. Good morning.
Leonov: Apollo, Soyuz. How do you read me?
Slayton: Alexey, I hear you excellently. How do you read me?
Leonov: I read you loud and clear.
Slayton: Good.

At 11:10 a.m. the spacecraft linked. Even before the space travellers could complete the docking procedures necessary to facilitate crew exchanges Brezhnev issued a statement to the Soviet and American astronauts,

> Speaking on behalf of the Soviet people, and for myself, I congratulate you on this memorable event....The whole world is watching with rapt attention and admiration your joint activities in fulfillment of the complicated program of scientific experiments. The successful docking had confirmed the correctness of the technical decisions developed and realized by means of cooperative friendship between the Soviet and American scientists and cosmonauts.

Finally at 2:17 p.m. Stafford entered the Soyuz craft, welcomed by a handwritten sign saying "Welcome aboard Soyuz." Shortly thereafter President Ford spoke with all five astronauts, congratulating them for their efforts and noting that it had "taken many years to open this door to useful cooperation in space between our two countries."[11] The three American astronauts and two Soviet cosmonauts then exchanged flags and other symbolic gifts..

The Apollo and Soyuz craft remained docked together for two days before going their separate ways (*Soyuz 19* landed on July 21, the Apollo craft on July 24). Overall, the joint mission was a veritable public relations blitzkrieg, and both nations' leaders reaped all the image benefits they could. The Soviet Soyuz launch was the first ever televised live in the United States; American guests at the launch included Ambassador Stoessel. Back in Washington, President Ford, NASA Administrator Fletcher and Soviet Ambassador Dobrynin watched the Apollo launch together live from the State Department auditorium. Topping off Brezhnev and Ford's personal statement and conversation, President Ford hosted all five crew members at the White House a few days later and declared, "This has opened a new era of international cooperation" which will hopefully "provide all of us with an example to remember for many, many years to come."[12]

Still, the 1975 effort was a limited one. From the beginning, the astronauts and cosmonauts were only intended to "visit both spacecraft and perform a number of simple scientific tasks."[13] Apollo-Soyuz was largely a public image meeting in orbit, a politically motivated project carried out by the two nations' space programs. Cooperation was largely limited to exchanging signatures and mementos--no valuable exchange of science was conducted in orbit.[14] Only five joint experiments were conducted,[15] and each country relied on its own communications links.[16]

This absence of scientifically productive joint experimentation was largely due to the fact that both nations were extremely wary of unwanted technology transfers.[17] Glynn Lunney, who was the American scientific project manager for ASTP, discussed this issue after the mission had taken place, defining technology transfers as beginning with discussions and descriptions of materials, systems engineering, management techniques and methods of training people to construct hardware. These, he said, would constitute the beginning of technology transfers, but, he explained, no such initiation ever took place:

> This is something we do not do. We are very conscious of our responsibilities and consider this to be a very important governing parameter. We keep firmly in our mind anything have [sic] to do with what we consider technology transfer.[18]

For example, despite the unprecedented spirit of cooperation, the U.S. refused to transfer equipment or designs of television cameras and fireproof spacesuit materials which the Soviets required for the unprecedented docking and the American oxygen-rich (and hence highly combustible) atmosphere, respectively.[19] Similarly, the United States built the docking mechanism (it was the same system Apollo used to dock with the lunar module) and the intermediate air-lock chamber, and the American astronauts did most of the work during the actual rendezvous and docking.[20] In addition, the ASTP involved older spacecraft: it was the Apollo craft's last flight, and the Soyuz had been used since 1968.

The Implications of the Apollo-Soyuz Linkup

Clearly there was a lot of show, but what is the legacy of the Apollo-Soyuz Test Project? The Apollo-Soyuz cooperation took place because both the Americans and the Soviets decided it benefitted them; neither side pursued cooperation for its own sake. The unilateral benefits ASTP bestowed upon the United States were threefold. First, although Soviet security agents carefully monitored American delegates in the USSR,[21] the cooperative project did enable the Americans to get a first-hand look at many of the Soviet procedures and equipment.[22] Second, the Apollo-Soyuz mission served as an important bridge between the now-ending Apollo era and the Space Shuttle era. The U.S. space program was about to face a major gap in its piloted space program. Without the Apollo-Soyuz linkup, there would have been no American manned flight between December 7, 1972 (the *Apollo 17* launch) and April 12, 1981 (the flight of the first Space Shuttle, *Columbia*). This gap necessitated NASA and corporate layoffs and inevitably damaged America's aerospace industry; the Apollo mission, as per the President Nixon's desire, ameliorated the effects of this gap.[23] Third, the Apollo mission continued for seven days after the docking, and allowed the astronauts to conduct twenty-three unilateral experiments, primarily the testing of new Earth science photographic equipment. Their experiments' results provided a "plethora of excellent photographic and observational data" for scientists from the United States, Australia, Barbados, Brazil, Canada, Egypt, India, Israel, Qatar and Turkey.[24]

The unilateral benefits to the USSR were quite different. The *Soyuz 19* mission did not fill in a convenient gap for the Soviets. Quite to the contrary, for the Soviets the mission came at a very inopportune time. From 1974 to 1976 the Soviets launched eleven Soyuz spacecraft and three Salyut space stations. One of these Soyuz missions ended in disaster for the Soviets. On April 5, 1975 the USSR launched a Soyuz spacecraft scheduled to dock with *Salyut 4*, but the mission was aborted when the Soyuz began tumbling violently at an altitude of approximately ninety miles. This mission failure threw off what was already a tight schedule, and the Soviets decided to take advantage of the perishables already aboard *Salyut 4* and try again with a second *Soyuz 18* on May 24, despite the fact that this necessitated having the *Soyuz 18* and *Soyuz 19* crews aloft at the same time. This simultaneity essentially doubled the ground support required. The logistical problems thus created were exacerbated by the fact that the Soviets were, during 1975, preparing to transfer their main ground control facilities from the Crimea to Moscow. Such was the scramble that some American Congressmen demanded that the Apollo-Soyuz Test Project be delayed to allow the Soviets time to adjust.[25] In the end no delays were required.

However, the Apollo-Soyuz Test Project did allow the Soviets, like their American counterparts, to gain a first-hand view of the American space program. The main benefit to the Soviets, however, was one of image. The Soviet image had not yet healed from the wounds of the moon race lost. Although publicly the Soviets insisted that they had never entered the manned moon race, the fact remained that the world knew that it was the American astronauts who had landed on the moon not once, but on six separate occasions. Below this public failure simmered the wounded pride of those Soviet scientists and engineers who knew all too well that they had raced and lost, wounds undoubtedly deepened by the fact that their government swept their now-abandoned efforts into a secret corner of shame. As Academician Keldysh revealed to an American scientist, Paul Doty of Harvard University, in the wake of the *Apollo 17* landing, "Our generation has failed." Dr. Doty showed a gathering of Academy scientists films of the America lunar landings, and these films brought forth "emotional cries of anguish."[26] It was in pursuit of equality of reputation that Brezhnev exaggerated the astronauts' and cosmonauts' "complicated program of scientific experiments," suggesting that the

Soviets and Americans were actually sharing what was, by implication, roughly comparable levels of skill and knowledge. The Apollo-Soyuz mission was an important salve for the Soviets' pride, as it gave the appearance of cooperation among equals.

The Apollo-Soyuz mission also benefited the Soviets and the Americans in a common fashion. Both the Soviet Union and the United States portrayed themselves as pushing aside national rivalries to cooperate in space for all mankind's benefit, just as the United Nations had originally urged in Resolution 1472 of December 1959. Apollo-Soyuz was advertised as the pursuit of science in its highest form by the world's dominant powers. Thus, the U.S. and USSR pursued a common image-related goal, aimed at their own citizens as well as at the peoples of other nations. This was an important consequence of the Apollo-Soyuz mission: it marked the first time Americans and Soviets had a vested interest in a joint space success. The complicated confluence of interests affecting both nations' space programs had overlapped in a significant fashion.

Plans for Post-ASTP Cooperation

According to each country's space leaders, the Apollo-Soyuz linkup was to be just the first in a line of further cooperative steps. The successful docking and modest scientific exchanges, combined with the operational and political barriers overcome, were described as preparation for further cooperation. The 1972 agreement had a five-year term, and ASTP was seen as an initial test of "compatible rendezvous and docking systems."[27]

Both Soviet and American leaders said that this mission was just the beginning of more substantive cooperation in space. Even as the two craft docked Brezhnev told the astronauts, "One can say that the Soyuz Apollo is a forerunner of future international orbital stations."[28] Shortly after the signing of the 1972 agreement, Academician Petrov stated, "I don't think that our joint effort will cease after the first experimental test flight....and I hope, very much, I'm sure that this first test flight will not be the last one--that there will be other similar flights...."[29]

During this same summer of 1972, NASA's George Low wrote Academician Keldysh requesting that in their May meetings they "outline the status of our thinking here with regard to future possibilities for cooperation." Low proposed cooperation in manned space flight (including joint Shuttle-Salyut missions and studies of future joint space stations), unmanned scientific missions (including a lunar farside sample return mission) and space applications (such as environmental monitoring and search and rescue).[30] From the outset Fletcher reflected and expressed the hope that the ASTP was just the beginning, stating, "All of us are quite optimistic that this new deeper cooperation in the exploration of space may lead to increased cooperation on still other programs."[31]

In pursuit of this increased cooperation the U.S. and USSR, from 1975 to 1977, negotiated a new space cooperation treaty to take effect once the five-year term of the 1972 agreement expired. On May 11, 1977, thirteen days before the older treaty was due to expire, the Americans and Soviets concluded the "Agreement Between the USSR Academy of Sciences and the National Aeronautics and Space Administration of the USA on Cooperation in the Area of Manned Space Flight."[32] The first part of the agreement established two joint working groups to study "the objectives, feasibility and means of carrying out a joint experimental program using the Soyuz/Salyut and Shuttle spacecraft," and gave 1981 as the target date for the first joint mission. The second part of the agreement, although rather pointedly sounding a realistic note ("Both sides recognize that no commitments are made at this stage concerning the realization of any project for creating an international space platform."), established a similar joint task force to study the possibilities of U.S.-Soviet cooperation in establishing a space station. In November 1977 American and Soviet scientists added a scientific and technical agenda to the agreement.[33]

Soviet officials welcomed the 1977 agreement with renewed predictions of increased collaboration. Petrov remarked in a *Pravda* interview that

> This agreement envisages the further development of cooperation in scientific and applied sphere of cosmonautics between the two countries. In particular, joint research will continued to be carried out in the space, near to earth, the moon and the planets and in the sphere of space biology and medicine, space meteorology, the study of the environment and creation of satellite search and rescue systems. The sides will adopt the necessary measures to further develop cooperation in the sphere of manned space flights....the pooling of Soviet and U.S. efforts in this sphere is a logical continuation of the successful 'Soyuz' and 'Apollo' flight.[34]

Cosmonaut/space program spokesman Herman Titov expressed even grander plans, suggesting that the U.S. and USSR work towards joint manned missions to Mars, saying that the enormous expenses involved "can hardly be defrayed by one country."[35]

Apollo-Soyuz had set the stage, and now perhaps the two superpowers might finally conduct truly cooperative ventures in space. All previous cooperative missions, including ASTP, could have been conducted unilaterally by either nation. The 1977 agreement suggested, although reservedly, the possibility that the United States and the Soviet Union might actually rely upon each other to expand the horizons of space in ways that they could not do (at least not as quickly or efficiently) on their own.

Shift in the International Climate

However, the 1977 agreement was ill-fated. Once again cooperation in space proved itself subject to the broader state of U.S.-Soviet relations. When these relations turned too sour, cooperation in an area so close to the increasingly vital military space programs evaporated. Shortly after the 1977 agreement, the Carter administration, which was determined to elevate human rights to the top of the American foreign policy agenda, became increasingly displeased with the Soviet human rights record.[36] Soviet suppression of the Polish Solidarity movement was condemned. Moreover, the U.S. and USSR became more heavily involved in conflicts over Ethiopia (a pro-Soviet Ethiopian government emerged and, with Soviet aid, crushed the U.S.-supported resistance). Further tensions mounted in U.S.-Soviet relations over Angola, Shaba, Yemen, Cambodia and Cuba. United States concerns were further heightened by the overthrow of the pro-Western Shah of Iran (February 11, 1979) and the seizure of American hostages by Iranian terrorists (November 4, 1979). Meanwhile Carter's hopes for achieving ban of anti-satellite weaponry were dashed when the Soviets insisted that the United States discontinue its space shuttle program as a prerequisite for any ASAT agreement.[37] The Americans could not have possibly entertained this demand: the shuttle was only months away from its first test flight and America's space program was counting on the shuttle to become its main launch vehicle. Although such tension and confrontations were hardly rare during the Cold War, the late 1970s saw a frequency of superpower conflicts which boded poorly for détente.

Thus, when the Soviet Union invaded (or, depending on who you believe, intervened at the request of the Afghan government)[38] Afghanistan on December 27, 1979, U.S.-Soviet relations entered a deep freeze. The next day Carter sent what he described as "the sharpest message" of his Presidency to Brezhnev over the superpower hotline, threatening that "unless you draw back from your present course of action, this will inevitably jeopardize the course of United States-Soviet relations throughout the world."[39] On January 5 an emergency United Nations Security Council meeting was called, where the Soviets vetoed an American-led condemnation of the intervention. In further actions, the Carter administration temporarily recalled the U.S. Ambassador to the Soviet Union, announced a boycott of the upcoming Moscow Olympics, requested a suspension in the Senate consideration of the SALT II Treaty, delayed opening new consulates-general in Kiev and New York, dismissed seventeen Soviet diplomats, recalled seven American diplomats, discontinued most Soviet and American cultural and economic exchanges, sharply restricted Soviet fishing rights in American waters, suspended all outstanding export licenses to the Soviet Union, tightened criteria for Russian nationals' travel to the United States and immediately put into effect an embargo on high technology items, phosphate and, most importantly, grain (a huge 17-million ton shipment was immediately cancelled; Chapter Five, pages 95-96 discusses Soviet agriculture during this period).

The Carter response was more than punitive. The United States also took proactive steps to strengthen its geopolitical posture and reach. Mandatory draft registration was reinstated, the U.S. navy intensified a massive buildup in the Indian Ocean, and began aggressive military maneuvers.[40]   Carter sped up the development of the new rapid deployment force, increased his request for the fiscal year 1981 military budget to $154.5 billion and announced a new $400 million aid package to nearby Pakistan.   An increasingly military-minded Congress then concluded a defense budget higher than the President's request--the first time this happened in U.S. history--allocating $159.7 billion. The increase of $28.7 billion over the fiscal year 1980 defense budget was the biggest rise in American peacetime history.[41]   On January 23 in his State of the Union message Carter defined what became known as the Carter Doctrine, declaring that any attempt "to gain control of the Persian Gulf region will be regarded as an assault on the vital interests of the United States of America, and such an assault will be repelled by any means necessary, including military force."[42]   As to when relations could warm up, Secretary of State Vance said on March 3, "Let me affirm today that the sanctions we have undertaken in response to the Soviet invasion will remain in force until all Soviet troops are withdrawn from Afghanistan."[43]

The Soviet troops did not begin leaving Afghanistan until May 15, 1988, and it was indeed almost that long before U.S.-Soviet relations improved. On November 4, 1980, Republican Ronald Reagan, riding a "get-tough" anti-Soviet platform, was elected by a large margin.[44]   Cooperation in space was impossible under these circumstances; once again, such joint work was subject to overall superpower relations.

The Legacy of the Apollo-Soyuz Mission

When astronauts Stafford, Brand and Slayton and cosmonauts Leonov and Kubasov shook hands in space, they pushed back not the frontier of science but the political barriers which had previously prevented cooperation at this level. Although cooperation in space remained subject to the international climate and more significant cooperation would be long in the coming, it would be wrong to underestimate the importance of the Apollo-Soyuz linkup. Though its practical accomplishments were small, the mission is important because it provided the precedent for top-level U.S.-Soviet cooperative initiatives in space. For the first time American and Soviet interests in space were intimately intertwined. ASTP showed that in an era of warmed relations, the two nations could work together in the cosmos, and it represented the successful culmination of unprecedented U.S.-Soviet exchanges at all levels of the political and scientific bureaucracies. These lessons would prove relevant for the future cooperative efforts described in Chapter Five.

It would have been unrealistic to have expected more: the facts remained in 1975 that the Soviet Union and the United States were superpowers competing the around the world for influence and that détente was relatively new and of questionable staying power. For more sustained and more meaningful collaboration, a better international environment would be required.

Notes

1. John F. Kennedy, "An Interview with John F. Kennedy," *Bulletin of the Atomic Scientists* vol. 16, no. 9, Nov. 1960: 347.

2. Article 1, "Agreement Concerning Cooperation in the Exploration and Use of Outer Space for Peaceful Purposes." A text of the agreement can be found in "Texts of the U.S.-Soviet Agreements on Cooperation in Space Exploration and in Science and Technology," *New York Times* 25 May 1972: A14.

3. This agreement also spawned cooperative ventures in medicine (including joint cancer, heart disease and arthritis research), seismography, chemistry, oceanography, air pollution control, semi-conductors and icebreaker and railroad technologies. Jonathan Spivak, "U.S.-Soviet Cooperation Yields Payload of Down-to-Earth Scientific Benefits" *Wall Street Journal* 18 Jul. 1975: 1, 23.

4. "Scientist Tells of USSR-U.S. Cooperation in Space Study," Moscow *Moscow Domestic Service* (30 May 1972) Translation by the Foreign Broadcast Information Service, *FBIS Daily Report--Soviet Union* vol. 3: 106; 31 May 1972: 2-3. Hereafter *FBIS*.

5. John J. Pomeroy and Norman J. Hubbard, *The Soviet-American Conference on Cosmochemistry of the Moon and Planets*, pt. 2 (Washington: National Aeronautics and Space Administration Scientific and Technical Information Office, 1977) iii.

6. The original formal discussions regarding the Apollo-Soyuz linkup began with the first ever NASA delegation visit to Moscow, which took place from October 26-28, 1970. During the trip, which was first suggested in a July 31, 1970 letter from NASA Administrator Paine to Academician Keldysh, the President of the Soviet Academy of Sciences, American and Soviet space officials discussed the feasibility of a joint docking. Glynn S. Lunney, "Trip Report - Delegation to Moscow to Discuss Possible Compatibility in Docking," 5 Nov. 1970. NHO 20.

Lunney was Chief of the Flight Director's Office at the NASA Manned Spacecraft Center in Houston.

7. Jeff Groman, *NASA* (London: Multimedia Publications, 1986) 84-85.

8. Arnold Frutkin, personal interview, 28 Jan. 1994.

9. Jonathan Spivak, "Ivan and John? The U.S. and Russia Seem Ready to Join Hands in Outer Space," *Wall Street Journal* 16 May 1972: 1, 23. Still, cooperation was not without its problems: even at this heightened level of U.S.-Soviet cooperation, not all contentious issues were resolved. For example, the President of the National Academy of Sciences noted in 1975 that, due to disapproval of Soviet treatment of dissidents and Jewish intellectuals, it was difficult to find "first-rate American experts" willing to cooperate with Soviet scientists. Robert C. Cowen, "East-West Science--the Uneasy Détente," *Christian Science Monitor* 2 Jul. 1975: 21.

10. On April 22, 1968 the U.S. and USSR signed the "Agreement on the Rescue and Return of Astronauts and Space Objects" and in 1973 they met at the Convention on International Liability for Damage Caused by Space Objects.

11. Ezell and Ezell, *Partnership* 327-31.

12. Linda T. Krug, *Presidential Perspectives on Space* (New York: Praeger, 1991) 111.

13. "Statement by Dr. Fletcher," 24 May 1972. James Fletcher was NASA Administrator at the time. NHO 7.

14. Leonid A. Vedeshin, personal interview, 13 Aug. 1993. Vedeshin, a member of the Soviet Academy of Sciences (and now of the Russian Academy of Sciences), worked at Interkosmos during the mid-1970s, where he helped manage ASTP's scientific projects. "There were some (fairly limited) experiments." Brian Harvey, *Race into Space* (New York: Halsted Press, 1988) 219.

15. Farouk El-Baz and D. M. Warner, *Apollo-Soyuz Test Project: Summary Science Report* vol. 2 (Washington: National Aeronautics and Space Administration Scientific and Technical Information Branch, 1979) iii.

16. McDougall, *Heavens* 431.

17. Kissinger was especially sensitive to technology transfers. In an August 1971 meeting with Acting NASA Administrator George Low, Arnold Frutkin and other officials, Kissinger emphasized that any technology transfer with the United States' European allies would "not be large and would be essentially controllable" but instead that "what would be transferred and what is desired by the Europeans is systems engineering and systems management know-how."

In this same meeting, the memorandum of which was originally declared "eyes only," the American leaders discussed the desirability of a U.S.-Soviet linkup. Although the memorandum does not explicitly describe policy decisions regarding technology transfers vis a vis the Soviets, given Kissinger's position regarding transfers to America's allies, it is safe to state that this *realpolitik* policy maker made it clear that any cooperation with the Soviets would have to be carefully tailored to avoid unwittingly yielding any scientific or engineering know-how. The memo may not have explicitly summarized this policy simply because it was likely to have been obvious to those present. George Low, "Items of Interest," Memorandum to [the new NASA Administrator] James Fletcher, 12 Aug. 1971: 1-2, 7. NHO 17.

For a discussion of the Soviet attitudes towards technology transfers, see Oberg, *Red* 139-40, 143.

18. Jim Maloney, "NASA Official Gets 'a little mad' Over Joint Space Mission Claims" *The Houston Post* 7 Nov. 1977: A7.

19. The Soviets had agreed to adjust their atmosphere to match the Americans', but this involved greatly increasing the Soyuz's oxygen content. The Soviets had not developed fireproof materials sufficient for the associated danger, and had to do so independently for this one mission after the State Department denied the technology transfer. In a similar event, the Soviets were forced to independently develop new camera technology for the mission. Arnold Frutkin, personal interview, 28 Jan. 1994.

20. Ezell and Ezell, *Partnership* 326-29. Oberg, *Red* 140-42.

21. Oberg, *Red* 139-40.

22. Maloney, "Official" A7.

23. George Low, "Memorandum for the Record: Meeting with the President on January 5, 1972," 12 Jan. 1972. NHO 19.

24. This experiment was called the "Earth Observations and Photography Experiment." El-Baz and Warner, *Summary* 1.

25. Oberg, *Red* 136-37.

26. Paul M. Doty, personal interview, 14 Jan. 1994.

27. Article 3. "Texts of the U.S.-Soviet Agreements."

28. Ezell and Ezell, *Partnership* 329.

29. "U.S.-USSR Cooperation in Space Press Conference," News Conference in Houston, Texas 13 Jul. 1972. The main participants were George M. Low, NASA Deputy Administrator and Academician Petrov. NHO 31.

30. George Low, letter to the President of the Academy of Sciences of the USSR, 24 Mar. 1975. NHO 18.

31. "Statement" 24 May 1972. NHO 7.

32. The agreement was formally signed in Geneva on May 18, 1977.

33. "Soviet, U.S. Scientists Discuss Cooperation in Space," Moscow *Moscow Domestic Service* (18 Nov. 1977) Translation by FBIS, *FBIS Daily Report--Soviet Union* vol. 3: 223; 18 Nov. 1977: 6.

34. "Academician Discusses U.S.-Soviet Space Cooperation Plans," Moscow *Pravda* (27 May 1977) Translation by the Foreign Broadcast Information Service, *FBIS Daily Report--Soviet Union* vol. 3: 109; 7 Jun. 1977: 3-4.

35. "What Next After ASTP?" *Sputnik* (Helsinki, Finland) Apr. 1976: 6.

36. An example of the Carter administration's willingness to subordinate space cooperation to human rights concerns came in July 1978 when the U.S. cancelled a trip to the Soviet Union by presidential science advisor Frank Press to protest Soviet treatment of dissidents. The trip's agenda would have probably included a discussion of the (already stalled) potential shuttle/Salyut joint mission. "Washington Roundup," *Aviation Week and Space Technology* 17 Jul. 1978: 13.

37. Richard Burt, "Soviet Said to Ask for Space Shuttle Halt," *New York Times* 1 Jun. 1979: A6.

38. Garthoff, *Détente* 935, 937 and 949.

39. Jimmy Carter, *Keeping Faith: Memoirs of a President* (Toronto: Bantam Books, 1982) 472.

40. After the invasion the U.S. sent two aircraft carriers (and their escort ships) to relieve the two which had been dispatched to the Persian Gulf after the Iranian seizure of American hostages. Among the replacement vessels was the Navy's newest carrier, the nuclear-powered *Nimitz*. In March 1980 the Pentagon announced that it would send an additional seven ships to the Indian Ocean with enough weapons and supplies to support several Air Force fighter squadrons and a ten thousand-man Marine amphibious brigade. *Congress and the Nation* vol. 5 (Washington: Congressional Quarterly Service, 1981) 110. On January 21, 1980 American B-52 bombers undertook unusually confrontational maneuvers, flying over a Soviet naval task force in the Indian Ocean.

41. *Congress* vol. 5, 125-27, 162-63.

42. *Public Papers of the Presidents of the United States: Jimmy Carter, 1980-81* (Washington: GPO, 1981) 197.

43. Garthoff, *Détente* 946-57.

44. Reagan carried forty-four states, receiving 489 electoral votes and 50.7% of the popular vote while Carter carried six states and the District of Columbia, receiving 49 electoral votes and 41.0% of the electoral vote.

5

A Confluence of U.S. and Soviet/Russian Interests in Space: Cooperation as the Cold War Ends

Reagan Ascendant: Crusade Against Communism

An improved international environment facilitating greater cooperation in space did not appear for over five years. From the invasion of Afghanistan in 1979 to 1985 the U.S.-Soviet relationship was characterized by antagonistic competition.

Upon assuming office, Ronald Reagan immediately began implementing the "get tough" program he had championed during his campaign. As the new President described his view of détente on January 29, 1981,

> so far détente's been a one-way street that the Soviet Union has used to pursue its own aims....I know of no leader of the Soviet Union...that has not more than once repeated...that their goal must be the promotion of world revolution and a one-world Socialist or Communist state....Now, as long as they do that...I think when you do business with them, even at a détente, you keep that in mind.[1]

Reagan increased Carter's proposed military budget (which Carter, before leaving office, had already increased by $32.6 billion) by an additional $6.3 billion; from 1981 to 1986 the U.S. spent $1.638 trillion on defense--$191 billion more than Carter had suggested even after the invasion of Afghanistan.[2] This military buildup included the acceleration of the deployment of the B-1 bomber and an enhanced version of the MX nuclear missile, deployment of the new Pershing II nuclear missile in West Germany, expansion of the navy from a 12- to a 15-carrier force and the development of a new generation of binary munitions and a new anti-satellite guided missile. The buildup also included accelerated development of stealth technologies, such as the B-2 bomber, F-117 and stealth cruise missile. The Reagan administration also actively promoted pro-Western governments throughout Latin America (especially in Nicaragua, El Salvador, Columbia and Honduras), flexed American muscle through direct interventions in Lebanon, Grenada and Libya and forged ahead with the Strategic Defense Initiative (SDI, or "Star Wars") ballistic missile defense system as fast as Congress would allow.[3]

The new weapons and SDI formed the vanguard of Reagan's intention to reassert American strategic superiority, to engage the Soviets fully in the superpower battle and to restore American credibility at home and abroad. The United States, Reagan publicly lamented, had lost its direction. The cause of freedom suffered from this loss of direction, Reagan declared, and the Soviet "evil empire" was advancing communism all too successfully.[4] The new administration's rhetoric suggested more than containment; it suggested confrontation. In this sense the Cold War of the first half of the 1980s took on a tone reminiscent of Dulles' "rollback" policy.[5]

The increase in U.S.-Soviet tension which resulted from the ascendancy of Reagan's Cold War policies was further reflected and exacerbated by events in Poland and Soviet airspace. From 1980 to 1982 the Polish Solidarity movement, led by Lech Walesa, pushed for liberalization. In October 1982 the Polish Communist Party cracked down, banning the workers' movement. The United States loudly protested this anti-democratic suppression. Then, on September 1, 1983, the Soviet military shot down Korean Airlines flight #007 as it strayed over Soviet airspace in the Far East. Two hundred sixty-nine civilians were killed. The Soviets were quick to defend their action as a justifiable defense of their country from espionage; Marshal Ogarkov, the Soviet chief of staff, claimed to have proof that the plane was on a "deliberate, thoroughly planned intelligence operation...directed from certain centers...in the territory of the United States and Japan."[6] American leaders charged the Soviets with intentional, barbaric mass murder. On September 2 President Reagan took advantage of the situation to state his views regarding the Soviet system:

> While events in Afghanistan and elsewhere have left few illusions about the willingness of the Soviet Union to advance its interests through violence and intimidation, all of us had hoped that certain irreducible standards of civilized behavior, nonetheless, [remained]....What can we think of a regime that so broadly trumpets its vision of peace and global disarmament and yet so callously and quickly commits a terrorist act to sacrifice the lives of

innocent human beings? What could be said about Soviet credibility when they so flagrantly lie about such a heinous act?

The answer, Reagan gave on the next day, is that "While the Soviets accuse others of wanting to return to the cold war, it's they who have never left it behind."[7]

The Korean Airline incident marked a low point of U.S.-Soviet relations, as characterized by Andropov's reaction in a formal statement describing Reagan's philosophy of world politics as

> a militarist course that represents a serious threat to peace. Its essence is to try to ensure a dominating position in the world for the United States of America without regard for the interests of other states and peoples....If anyone had any illusions about the possibility of an evolution for the better in the policy of the present administration, recent events have dispelled them once and for all.[8]

Thus, the propaganda machines of both countries, led by Reagan and Andropov, clashed with great ferocity, entrenching themselves ever more deeply in what must have seemed to be the permafrost of the Cold War.

During this era of high tension, formal intergovernmental cooperation in space was impossible. Space was, from 1979 to 1985, principally an extension of the heightened arms race. Nuclear arms were the critical aspect of this arms race and, due to the growing capabilities and importance of military satellites and the rise of SDI, nuclear arms became more intertwined than ever with space. In 1982 publicly-known Department of Defense space spending exceeded NASA's budget for the first time since 1960.[9] Reagan, during his first five years in office, saw space not as an area for establishing common ground but as the high ground where the United States could use its technological and economic superiority to gain undeniable military preeminence and, it followed, unquestioned world leadership.

Gorbachev Ascendant

Superpower relations had turned sour in late 1979 and only got worse in the ensuing years. However, beginning in 1985 events began to change with increasing speed until there was an improved international climate which facilitated an era characterized by U.S.-Soviet/Russian space cooperation. The radical changes began in the Eastern Bloc.

In March 1985 Chernenko became the third Soviet First Secretary to die in as many years. Mikhail Gorbachev succeeded Chernenko, and inherited a Soviet state reeling from decades of stagnation under

Brezhnev and most recently from two leaders who had inherited their positions of power extremely ill (Andropov, First Secretary from November 1982 to February 1984, was a diabetic with a renal condition that required a kidney transplant during his tenure; Chernenko suffered from severe emphysema). Gorbachev, like Khrushchev some thirty years earlier, found himself leading a nation haunted by looming agricultural disaster.[10] Even more ominously, this time around the Soviet Union faced declining world power and general economic collapse. While Khrushchev believed the USSR was on the path to outpacing the United States economically, Gorbachev recognized that his country was on the brink of disaster.

Before 1985 Gorbachev had, as a regional First Secretary and Central Committee Secretary for Agriculture, gained a reputation as an innovative--but still loyally communist--leader. As First Secretary he led the nation through a series of reforms centering around the increasingly radical banners of *uskorenie* (acceleration), *perestroika* (rebuilding), *glasnost* (openness) and *demokratizatsiya* (democratization). Gorbachev loosened the reins of censorship, introduced limited forms of private ownership and tried to discipline and appease increasingly assertive nationalities in the hopes of revitalizing the decaying empire. Although Gorbachev failed to "save" the Soviet Union, perhaps his greatest domestic legacy is his introduction of the first truly competitive elections in the Soviet Union in the spring of 1989.

The Return of Cooperation

Gorbachev Challenges the U.S.'s Peaceful Mettle

Gorbachev's foreign policy legacy may well prove to be that of the Communist Party leader who oversaw the dismantling of the Iron Curtain. His policies of liberalization contributed to 1989 revolutions in Bulgaria, Czechoslovakia, East Germany, Romania, Hungary, Poland and March 1990 declarations of independence from Lithuania and Estonia; all of these events would have invited decisive Soviet intervention in years past. Gorbachev approved the opening of the Berlin Wall in November 1989 and the unification of Germany eleven months later (although it is not clear he could have stopped it). Well before an August 1991 failed conservative coup critically weakened Gorbachev's power base, the reformist First Secretary had allowed if not overseen the end of Soviet Communist hegemony in Eastern Europe.

Gorbachev also extended an olive branch to the United States, helping to lead the superpowers toward the negotiated end of the Cold War. As soon as Gorbachev took power in March 1985, the U.S. and USSR resumed arms negotiations, and from November 19 to 21 of that same year Reagan and Gorbachev met for the first time in Geneva. In the next three years the superpower leaders met three more times and signed the INF (Intermediate-range Nuclear Forces) treaty agreeing to eliminate an entire class of nuclear weapons.[11] The INF treaty was a critical precedent for space cooperation, since it established mutual on-site inspections. On May 15, 1988 the Soviet Union began withdrawing its troops from Afghanistan, meeting a precondition to a fundamental improvement of U.S.-Soviet relations that had existed since Cyrus Vance had insisted that American protests would not cease until the Soviets reversed their invasion (Chapter Four, page 88).

Then, in a December 1988 address to the United Nations, Gorbachev challenged the United States to join the Soviet Union in disarming and took the initiative, announcing the unilateral withdrawal of 50,000 troops and 5,000 tanks from Eastern Europe, the reduction of the Soviet armed forces personnel by 500,000 and the destruction of 10,000 tanks, 8,500 artillery systems and 800 aircraft. Economic considerations would have forced these reductions eventually but Gorbachev took the initiative to make a bold diplomatic proposal. Soon thereafter the Soviet Union loosened its position positions on the stalled Strategic Arms Reduction (START) and Conventional Forces in Europe negotiations. The superpower thaw continued to gain increasing momentum, progressing through the agreement to destroy chemical weapons (June 3, 1990), the conclusion of the Conventional Forces in Europe agreement (November 19, 1990) and the START 1 and 2 Treaties (July 31, 1991 and January 3, 1993, respectively).[12] Indeed, a veritable *dis*armament race began, as Presidents Bush and Gorbachev followed one another in announcing unilateral arms cuts in September and October 1991.[13]

The momentum of these improvements in U.S.-Soviet and then U.S.-Russian agreements led President Bush to conclude on January 2, 1993, that there was "a new U.S.-Russian partnership...affirming our dedication to democratic peace....The two powers that once divided the world have now come together to make it a better and safer place." Yeltsin echoed Bush's remarks the next day, describing the United States and Russia as "partners who not only trust each other but also assist each other."[14]

An important part of this assistance was the aid the United States had begun channeling towards its sworn enemy turned potential ally. Although the aid promised far outweighed the aid actually delivered, this assistance marks a significant evolution of Western policy vis a vis the former Soviet Union. The process began at the July 1991 Group of Seven (G-7) economic summit when the seven leading democracies (the United States, Japan, Germany, Canada, Great Britain, France and Italy) agreed to offer the Soviet Union technical aid and associate membership in the

International Monetary Fund (I.M.F.) and World Bank. However, the G-7 leaders stopped short of offering the USSR monetary aid. As relations improved the West offered Russia more and more direct aid, and, on April 1, 1992, President Bush announced a $24 billion aid package to Russia. Of this $24 billion some $15 billion was delivered by early 1994. At an April 1993 meeting of finance and foreign ministers, the G-7 expanded this first package, promising to deliver $43.4 billion. The aid included funding from the I.M.F., World Bank and the European Bank for Reconstruction and Development and $15 billion of debt relief. By February 1994 the G-7 had delivered the $15 billion of debt relief, but only $8 of the other $28.4 billion promised at the April 1993 conference.[15]

Space Cooperation Reborn

The improved international climate allowed the rebirth of significant U.S.-Soviet space collaboration. The first steps towards renewed space cooperation were taken in Moscow on April 15, 1987, when Secretary of State Shultz and Foreign Minister Shevardnadze signed an Agreement Concerning Cooperation in the Exploration and Use of Outer Space for Peaceful Purposes. The 1977 agreement on bilateral space cooperation had been allowed to expire in 1982; the 1987 accord renewed the drive towards joint efforts, resolving to "carry out cooperation in such fields of space science as solar system exploration, space astronomy and astrophysics, earth sciences, solar-terrestrial physics and space biology and medicine" and to "encourage international cooperation in the study of legal questions of mutual interest which may arise in the exploration and use of outer space for peaceful purposes."[16]

As a part of the agreement the space powers concluded a sixteen-item "agreed list of cooperative projects," including exchanges of data on the Venusian surface, cosmic dust, meteorites, lunar materials and radio, gamma-ray, x-ray, solar and sub-millimeter astronomy. Foreshadowing three future areas of major collaboration, the two parties agreed to a coordinated study of Mars landing sites, cooperative "activities in the study of global changes of the natural environment," and exchanges regarding space biology and medicine (to lead to the publication of a second edition of the *Principles of Space Biology and Medicine*).[17]

Although the agreement was modest in scope, emphasizing data exchanges and joint studies, it marked an important turning point in superpower space relations. The accord started the two nations on a path towards greater joint efforts in space. This progress notwithstanding, the fact that much remained to be done before greater cooperation would take place is evidenced by the way that President Reagan, in a statement summarizing Shultz's Moscow negotiations, failed to mention the accord or even the existence of space cooperation as a part of the meetings' agenda. Instead the president cautiously, if not cynically, noted in passing that "agreements on...space and defense will be...difficult."[18]

The Apollo-Soyuz Test Project Eclipsed

Cooperation in Earnest from 1987 to 1991

As superpower relations rapidly improved between 1987 and 1991, the two space powers became more amenable to increased collaboration. On August 15, 1991 (four days before the failed anti-Gorbachev coup began) the Soviet Union launched its atmospheric monitoring satellite, *Meteor 3*. *Meteor 3* carried NASA's Total Ozone Mapping Spectrometer, and was the first Soviet satellite to carry American instrumentation.[19] The *Meteor 3* satellite outlasted the USSR, and American and Russian scientists continued to collect data from the instrument through early 1995.[20] This was the largest U.S.-Soviet space hardware program since the Apollo-Soyuz mission, and provided an opportunity for the United States and Soviet Union to lay the foundation for further cooperation in a benign field.[21] During this same period the U.S.-Soviet Earth Science Working Group was established. This organization issued a joint ozone report every two to three months publicizing the latest data on the ozone layer, considered more launches of American instrumentation on Soviet rockets and studied the possibility of merging the Soviet and American civil weather broadcasting programs. In addition, in August 1991 the Soviet Union offered the United States access to its three biggest radio telescopes as a part of the Search for Extra-Terrestrial Intelligence program and in October 1991 Department of Defense and NASA officials traveled to the Soviet Union to consider expanded uses of Soviet launch vehicles.[22]

Bush and Yeltsin Sign Landmark Pact

On August 19, 1991, the evening before Gorbachev intended to sign a union treaty surrendering many of the Soviet Union's centralized controls over the republics, conservatives placed Gorbachev under house arrest and declared emergency powers. Despite the fact that their conservative coup failed within two days, in the weeks and months to come the Soviet Union seemed a dying horse begging to be shot.

The attempted coup began the end of Gorbachev's reign as the last Soviet leader. Within two month's of the failed coup all three Baltic states became independent of the Soviet Union and the republic of Georgia declared its independence. Meanwhile Boris Yeltsin, riding the wave of popularity begun with his March 4, 1990 election to parliament and elevated by his June 13, 1991 election as the first popularly elected president of Russia and his successful defiance of the conservative coup, began asserting Russia's political power in an unprecedented manner. On November 6, 1991 President Yeltsin issued a decree outlawing the Communist Party of the Soviet Union on Russian territory, directly challenging the heart of Gorbachev's power. No sovereign Soviet Union and Russia could coexist, and, on December 8, 1991, Yeltsin joined Ukraine's President Kravchuk and the Chairman of Belarus' Supreme Soviet, Shushkevich, in declaring the Soviet Union an entity of the past:

> Noting that the short-sighted policy of the center has led to a profound economic and political crisis, to the breakdown of production and to a catastrophic drop in the living standards of practically all strata of society, [b]earing in mind the growing social tension in many regions...[we] [p]roclaim the establishment of the Commonwealth of Independent States.

That same day in a separate protocol Yeltsin, Kravchuk and Shushkevich declared "that the Union of Soviet Socialist Republics as a subject of international law and a geopolitical reality no longer exists."[23] Thirteen days later, on December 21, 1991, Armenia, Azerbaijan, Kazakhstan, Kyrgyzstan, Moldova, Tajikistan, Turkmenistan and Uzbekistan joined the Commonwealth of Independent States (CIS).[24]

On Christmas Day 1991 Gorbachev resigned his post as President, acknowledging the dissolution of the Soviet Union. That night the red, white and blue Russian flag flew over the Kremlin. The world's largest country was no more, and Russia, the main inheritor of the Soviet space program (Chapter Seven), turned to the West even more for direct aid and overall cooperation.

The continuing improvement of U.S.-Russian relations gave further impetus to closer space cooperation. On June 17, 1992 Presidents Bush and Yeltsin signed an Agreement Concerning Cooperation in the Exploration and Use of Outer Space for Peaceful Purposes, calling for increased U.S.-Russian collaboration in space. Within two weeks of the Bush-Yeltsin summit, NASA gave the Russian aerospace firm Energia a one million dollar contract to study how the United States could use Russian space hardware (this contract was expandable to $10 million).[25] This agreement--the first between NASA and a Russian company--was a significant departure from the Cold War.[26] Suddenly the American space program was directly funding its historical competition.

In order to implement and expand upon the June 1992 pact, an American delegation led by the Administrator of NASA, Daniel Goldin, and the Executive Secretary of the U.S.'s National Space Council, Brian Dailey, traveled to Russia for discussions with the Russian Space Agency (directed by Yuri Koptev), Space Research Institute (IKI, directed by Academician Galeev), the Russian Academy of Sciences and the Babakin Research and Test Center of the NPO (Scientific Production Association) Lavochkin.[27] The leaders of the two nations' space programs initiated several new joint ventures, including the expansion of the Working Group on Space Biology and Medicine (established in the 1987 agreement), the formation of a working group to oversee bilateral cooperation regarding NASA's environmentally oriented "Mission to Planet Earth," and the inception of a joint study of "enhanced cooperation in monitoring the global environment from space." Also examined was mutual tracking of the American planetary probe *Galileo*. These discussions were based on a 1992 contract between NASA and the Russian Institute of Space Device Engineering to study the use of the Russian Deep Space Communications Network to augment the NASA Deep Space Network antennas.

Acknowledging the importance of establishing direct interpersonal links, the Goldin/Dailey-Koptev conference also arranged for a 45-member delegation of U.S. government and industry officials to visit Russian space-related enterprises. The American delegation, which was headed by the Director of the Office of Space Commerce and included senior management and engineers from seventeen leading aerospace firms, visited over forty Russian design bureaus, scientific production associations, research institutes and production enterprises from July 18 to July 29, 1992. At least four of the American companies signed contracts with Russian space interests within a year of their visit, pavinghe ways for expanded commercial space cooperation.[28]

Cooperation in Unmanned Missions to Mars

The Goldin/Dailey-Koptev conference also began what became the rapidly arranged American involvement with the Russian Mars 94/96 missions. The studies begun during the July 1992 conference led to an October 1992 inter-agency agreement. In a flurry of activity, NASA overcame what the Russians described as deadlines already past,[29] starting with a $2 million contract with the Russian Space Agency (RKA)--the first contract between the two countries' civilian space program leaders--to study the possibilities of developing a Russian-U.S. hard-lander for drop onto Mars from the Russian/French Mars 94 orbiter. The contract also examined the possibility of piggy-backing a NASA-Jet Propulsion Laboratory (JPL) micro-rover on the larger Russian rover for the more ambitious Mars 96 mission. In pursuit of these goals teams from the IKI, Russian Academy of Sciences, Babakin Research and Test Center, American ISX Corporation, NASA and the International Planetary

Society regularly flew between the countries, testing rovers in Kamchatka and the deserts of California.[30]

The NASA contribution surpassed the original plans and, as a result of a $1.5 million NASA/JPL-Russian contract, the Mars 94 mission was intended to carry an American-led Mars Oxidant Experiment (MOX) to the Martian surface, where it was to collect spectrometry, seismological, optical and oxidation data. MOX included instruments from Russia, France, Germany and Finland--a novel and truly multinational effort designed take advantage of the rare landing. An unprecedented nineteen nations (Russia, the United States, Austria, Belgium, Bulgaria, Finland, France, Germany, Great Britain, Greece, Hungary, Ireland, Kazakhstan, Italy, Latvia, Poland, Sweden, Switzerland and Ukraine) were a part of the Mars 94 mission. This multinational technological marriage was motivated by scientific goals and skills: each of the nineteen nations was contributing what it did best[31] to a Russian-initiated project which sought cooperation after running out of money.[32] Mars 94/96 was an example of cooperation which was primarily scientifically motivated and promised significant scientific results. In addition, the U.S.-Russian collaboration plans spanned three chronological periods: short-term (Mars 94) intermediate-term (Mars 96) and long-term (the two countries are exploring possibilities for a human mission to Mars).[33] Thus, planning for Mars collaboration, facilitated by the improved international climate, had, by early 1994, already gone beyond that associated with the Apollo-Soyuz project: instead of constituting a one-time arms-length public relations meeting in space, this cooperative project sought repeated exchanges of technology and promised to yield otherwise unattainable scientific results.

However, these grand plans never came to fruition. Exploration of Mars suffered a major blow when NASA lost contact with its multi-billion dollar Mars Observer probe. Mars Observer was intended to, among other things, transmit more detailed information about Mars, thus helping scientists pick landing sites for future missions. The probe was also going to serve as a data relay link for future missions. Thus, the Observer's failure made future missions to Mars more difficult.

Although the Mars collaboration reflected the more friendly international climate of the early 1990s, it also evidenced Russia's economic woes. As of April 1993, the Mars 94 launch was on schedule for an October 1994 launch, with arrival at Mars planned for September 1995.[34] However, toward the end of 1993, the Russian Space Agency announced that it could not launch Mars 94 unless its Western partners could provide $10 million for the mission.[35] The Western partners (mainly the U.S., France and Germany) did not meet this last-minute request, and the Mars 96 mission was pushed back to 1998 (becoming

Mars 98). Soon thereafter Mars 94 was delayed to 1996 (becoming Mars 96).[36]

This unilateral withdrawal angered Russia's Mars collaborators. Western officials had long suspected trouble, but, although they had repeatedly asked their Russian counterparts if delays were imminent, the Russians waited until the last minute to provide their fait accompli: come up with millions of dollars or no mission. As similar tension continued, Germany considered cancelling a proposed (paid) 1996 visit to Mir. Germany and France had already invested approximately $200 million in the collaborative Mars missions, and by early 1995 it was still unclear what version of Mars 96 the Russians would actually launch.[37]

Mars cooperation only got more complicated. The United States reverted to its own missions, the relatively inexpensive Mars Pathfinder for launch in December 1996, the Mars Global Surveyor for launch in November 1996, and the two (separate) Mars Surveyor missions for launch in late 1998.[38] In the summer of 1994 the United States began exploring the possibility of merging the second of its Mars Surveyor missions with Russia's Mars 98 program, but it was clear that NASA was working to ensure that Mars Surveyor would go forth, with or without the Russians. There were also discussions of merging exploration programs for the 2001 launch window.[39] Planning for the potential "Mars Together" project revolved around mixing primarily American, Russian, French and German hardware. However, neither the Americans nor the Russians had much time for the negotiations, since they hoped to include any agreements in their fall 1994 budget requests for the fiscal year 1996.

The Mars Together initiative was problematic from the outset, because during the spring of 1994 Russia's Space Research Institute announced it could no longer afford a Proton for the 1998 launch, and would instead try to use the smaller Molniya booster. The Molniya would not be able to carry nearly as much equipment to Mars as the Proton or the American Med-Lite. (The American Med-Lite was planned for use if Mars Surveyor 1998's first launch remain a solely American endeavor.)

The U.S.-Russian Mars Together negotiations dragged on until November 1994, when NASA Administrator Dan Goldin asked his counterpart, Russian Space Agency Director Yuri Koptev, for a commitment. Goldin continued to push for a Proton launch vehicle. Koptev responded that Russia could not afford it.[40] Although the books on Mars Together were not officially closed until May 1995, hopes for a U.S.-Russian mission to Mars before the end of the 20th century effectively died that day.

Although this failure may have boded poorly, it was far from fatal for U.S.-Russian space collaboration. In fact, even as the joint Mars missions were falling to pieces, American and Russian officials began discussing collaborative missions to launch probes to Pluto and toward the Sun.[41] The proposed projects, "Fire and Ice," were originally JPL missions tentatively planned for early in the 21st century. Discussions began in earnest in April 1994 when U.S. and Russian Technical Working Groups met to discuss the missions. The motivations for collaboration in this case are reflective of the times: American backers of the program hoped not only to gain access to relatively inexpensive Russian boosters, but also to increase their program's chances of being funded by tying them to Russian collaboration; the Russians, for their part, were interested largely because the projects would give them unprecedented access to American small-probe technologies.[42]

The Pinnacle of Cooperation in Space

Mars cooperation, despite an auspicious beginning, proved to be an example of the difficulties of Western-Russian space collaboration. However, even as Mars collaboration decayed, joint space work in another realm became highly successful. The July Goldin/Dailey-Koptev meeting also gave birth to what would become, as of early 1994, the pinnacle of U.S.-Russian cooperation in space. Citing the goals of "maximization of scientific return, development of rescue capabilities, and mutual understanding of respective engineering, safety, and training approaches," the 1992 Bush-Yeltsin agreement outlined three phases of increasing collaboration involving the first joint manned missions since 1975. The first phase involved the flight of a Russian cosmonaut aboard the Space Shuttle, the second the flight of American astronauts aboard the Russian space station Mir and the third rendezvous, docking and joint missions between the Space Shuttle and Mir.

In order to accomplish these missions NASA and the Russian Space Agency signed a Memorandum of Understanding in June 1994. In this contract the goals of the Shuttle-Mir missions were outlined as follows:

- Allowing American astronauts to spend up to 24 months on board Mir, giving the U.S. valuable experience in long-duration stays in space.
- Using the space shuttle to deliver people and cargo to Mir. This would provide experience for the two sides in working together, i.e., in command-and-control and risk reduction.
- Conducting joint scientific experiments on Mir.
- Using Russian equipment for docking.
- Co-developing advanced solar power systems.
- Advancing extravehicular activity (space walk) techniques.
- Testing technologies for construction of space station Alpha.
- Giving Russian scientists up to $20 million for relevant ground-based research.

• Funding initial development of the first space station module, "FGB."[43]

These accords were rapidly implemented. On February 3, 1994 the U.S. launched STS-60, the Space Shuttle *Discovery*, carrying Sergei K. Krikalev, the first Russian cosmonaut launched on an American spacecraft. Krikalev was the first foreign astronaut to undergo full NASA training as a mission specialist (all previous guest astronauts were payload specialists) and took part in seven experiments. Vladimir G. Titov, who trained with Krikalev, was the second cosmonaut on a shuttle mission: he was launched on space shuttle *Discovery*, on February 3, 1995, exactly one year after his comrade.[44] On this mission the space shuttle flew within 37 feet of Mir in order to prepare for future dockings.

Shortly thereafter American astronauts had their first experiences as guest cosmonauts. On March 14, 1995 Norman E. Thagard became the first American astronaut launched on a Russian spacecraft, flying aboard a Baikonur-launched Soyuz to space station Mir for what was originally scheduled as a three-month stay. Thagard's otherwise successful mission was somewhat marred by Russian delays in delivering equipment he was intended to use for experiments. The Space Shuttle *Atlantis* was specially outfitted for docking with Mir, and made the first shuttle trip to Mir on June 27, 1995, docking with the Russian station on June 29. Together the shuttle and space station held a record ten people, and the enthusiastic astronaut-cosmonaut meeting was highly reminiscent of the Apollo-Soyuz docking twenty years earlier. *Atlantis'* trip to Mir was to be the first of seven shuttle missions to Mir, and Thagard's one hundred-plus day stay broke the previous American record of 84 days in space.[45] The remaining six shuttle docking missions were scheduled for October 1995, March 1996, August 1996, December 1996, May 1997 and September 1997.

Prelude to a Joint Platform

The July 1992 Goldin/Dailey-Koptev meeting also stimulated discussions of the use of the Soyuz as an emergency Assured Crew Return Vehicle for space station Freedom and the use of the massive Soviet Energia rocket to launch parts of Freedom. These projects alone would have constituted a major departure from Space Station Freedom's original purpose as a highly visible symbol of Western strength. In announcing his decision to direct NASA to develop the space station on January 25, 1984, President Reagan described the project as a reflection of the fact that "We are first; we are the best; and we are so because we're free."[46] Vice President Quayle echoed this motivation in 1990 when he wrote, "Science is but one reason for building a space station....The ultimate mission of the Space Station is...the reaffirmation of the leadership in space of the United States of America, the world's only superpower."[47]

However, the scope of the U.S.-Russian cooperative discussions continued to expand rapidly. U.S. and Russian space officials came to see the joint manned missions as laying the foundation for what would comprise the first U.S.-Russian space project which involves cooperation from start to finish, from research and development to construction and use: Space Station Alpha.

Space Station Freedom had come to face a rapidly intensifying crescendo of budgetary pressure in the early 1990s. NASA's overall budget was $13.9 billion in 1991, $14.0 billion in 1992 and $14.3 billion in 1993.[48] Given these increasingly tight budgets (none of them matched inflation's rise; these same figures are $12.2 billion, $11.7 billion and $11.6 billion in 1987 dollars), Freedom, as NASA's biggest project and expenditure, naturally became the prime target for cutbacks. In early 1993 the Congressional Budget Office placed space station Freedom at the top of its list of options for trimming discretionary domestic spending, estimating a potential savings of $10.4 billion over five years.[49] In mid-March 1993 President Clinton ordered NASA to downsize Freedom (the station's third major cutback), and, on June 17, 1993, he announced that the station's ten-year budget would be slashed by $18 billion. As he announced these cuts Clinton declared that he would

> seek to enhance and expand the opportunities for international participation in the space station project, so that the space station can serve as a model of nations coming together in peaceful cooperation.[50]

Space Station Freedom was already planned as a cooperative project between the United States, Japan and the European Space Agency; when President Clinton talked of enhancing and expanding joint efforts, he was referring to the world's leader in extended stays in space--Russia.[51] In fact, Russia, had already been planning to develop a second-generation Mir 2 space station when it agreed to join forces in developing Alpha.[52]

Presidents Clinton and Yeltsin agreed in principle to merge Mir 2 and Freedom's efforts during their April 1993 Vancouver summit. Then, on September 2, 1993, American Vice President Gore and Russian Prime Minister Chernomyrdin signed an accord agreeing to merge their nations' space station efforts. Mir 2 and Freedom were no more; the U.S. and Russia would, together with Freedom's European and Japanese partners, build space station Alpha. A remarkable event occurred in April 1994 in Washington, D.C. The heads of the world's biggest civil space programs--NASA, the Russian Space Agency, the Canadian Space Agency, the European Space Agency, and Japan's Space Development

Agency--gathered together in one place with a common goal: the international space station.

Space station construction was organized to take place in a three-phase schedule (not the same one as for the Shuttle-Mir missions). Under this plan Phase I included the joint piloted flights and was to be conducted from 1995 to 1997, Phase II included some seven Russian and four American flights dedicated to building the platform and was to occur from November 1997 to June 1998 and Phase III involved finishing the station from 1998 to 2002.[53] Construction would begin in November 1997 with the launch of the "FGB" module mentioned above. The Russian space enterprise Khrunichev was given a $190 million contract by Lockheed Martin, subcontractor to NASA's prime space station contractor, Boeing, to construct this first module. These dates had been revised significantly in November 1993 and February 1994, but did remain stable from March 1994 to at least August 1995.

The Americans looked to the cooperative space station most explicitly to cut costs. Less touted but also important was the enhanced sense of purpose which Russian cooperation gave the American space effort. The increasing budget pressure had created a need to enhance the project's vision. It was an era of budget cuts, and Congress had ominously proven its willingness to cut large science projects already in progress by killing the supercollider project. Space station Freedom had been widely criticized as floundering for a *raison d'etre* and was reeling from charges of mismanagement. Many scientists saw the project as a waste of scarce money, and a 1990 NASA study showed that the station's design was fundamentally flawed and would require a literally impossible amount of maintenance. As a result the expensive project faced increasing Congressional scrutiny.[54]

To survive another year's budgetary hearings the project needed presidential support. A few years before Quayle had lobbied Congress by emphasizing Freedom as a symbol of America's strength. However, in 1993 Congress was less concerned with further reaffirmations of America's power vis a vis the languishing remnants of the Soviet empire.

Meanwhile President Clinton stood to gain from a foreign policy success. Clinton had been widely criticized for handling international relations poorly (especially regarding the widening crises in Somalia, Haiti and Yugoslavia), and now Clinton seemed to risk being labeled as the American leader who had "lost Russia." In 1993 Russia's reforms marked a rocky, meandering path at best, and some would-be critics seemed all too ready to, should Russia revert to an anti-Western state, accuse Clinton of not having done enough to "save Russia." Even former President Nixon publicly pressed the new American president, writing in a *New York Times* op-ed piece that "it would be tragic if, at this critical point, the United States fails to provide the leadership only it can provide."[55]

It was in this highly charged environment that Clinton chose the joint space station as cooperation aimed at supporting Russia's fledgling democracy and intimately associated with the larger currents of overall U.S.-Russian relations. Space station Alpha was a highly visible aspect of U.S. policy towards Russia and was intended to highlight what administration officials described as Clinton's greatest foreign policy success.[56] As Goldin described Alpha, "There is no event that can better define the coming of the new age than we join with Russia and actually invest in technology instead of building weapons."[57] Thus, Clinton linked his political fortune to the space station. Given this linkage the joint space station's fate in Congress grew more certain; without presidential support Congress was much more likely to cut the project.

Russia looked to the cooperation out of financial expediency. Russia's space budgets had been slashed and Mir 2's fate seemed dark at best. In August 1993, a few weeks before Chernomyrdin and Gore concluded the space station agreement, Koptev described Russia's space program as on the verge of a financial ruin, saying that unless the government implemented "an emergency survival program for the space industry, the collapse could happen even during this year."[58] The Russian space program was especially desperate for hard currency, and this is what it secured in the September 1993 accord. As a part of the agreements NASA agreed to pay Russia $400 million, $305 of which was to be disbursed from 1994 to 1997 in return for the American astronauts' time on Mir. Russia was to spend the other $95 million on research and development for Space Station Alpha. Koptev described the situation succinctly: "We're getting out of a tough spot today."[59]

Other Aspects of Enlarged Cooperation

During their September 1993 meeting the Vice President and the Prime Minister also arranged for expanded cooperation in environmental monitoring, space science and aeronautics, as well as for Russian entry into the commercial launch market.[60] Russia also signed a bilateral agreement with the U.S. promising to adhere to the guidelines of the Missile Technology Control Regime (MTCR).

The Clinton administration had set adherence to the MTCR as a prerequisite for Shuttle-Mir and space station cooperation agreement, and for allowing Russia to launch commercial satellites. The Missile Technology Control Regime took effect on April 16, 1987, and originated from an agreement between the United States, Canada, France, West Germany, Italy, Japan and Great Britain intended to "limit the proliferation of missiles capable of delivering nuclear weapons." The United States requested that Russia formally conform to the agreement's principles and guidelines to prevent the sale of technologies which could support another nation's military rocket programs.

In addition, NASA planned to contribute new sensors, research gear and capsule environment equipment intended to enhance the capabilities and extend the lifetime of Mir. American and Russian specialists also began working together in late 1993 to develop new solar power, environmental control, life support and common space suit systems.

Taking A Deep Breath: The Rapid Changes from 1991 to 1995

Thus, the end of the Cold War created a political environment which allowed productive, unprecedented U.S.-Russian cooperation in space. The *Meteor 3* launch, NASA contracts with the Russian Space Agency and NPO Energia and studies of additional potential areas for Russo-American space cooperation (e.g., studies on the Assured Crew Return Vehicle, Energia launches for space station Freedom assembly and an international space platform) ushered in this new era of renewed collaboration. International Mars missions showed initial promise but then fell victim to Russian budget constraints and the partners' inability to find common ground. Discussions of joint missions to the Sun and Pluto picked up where Mars collaboration went wrong, hoping to do better. Finally, most significantly, the United States and Russia essentially merged their human space programs, including their space station efforts. Vice President Gore, in announcing the September 1993 accord, acknowledged the importance of the Apollo-Soyuz Test Project precedent and touted the primary selling point of joint work in space:

> the agreements that we signed here today...most clearly have their roots in the Apollo-Soyuz rendezvous and docking in July 1975. It was through this project that Russian and American space scientists and engineers, astronauts and cosmonauts first began to work together....The future holds more of what the Apollo-Soyuz project foretold: *close work together to minimize the costs and cut the time needed to do the projects while achieving more than would otherwise have been possible* (emphasis added).[61]

What, however, was the Vice President using as his frame of reference for determining that cooperation would yield more efficient exploration and exploitation of outer space? "More than would otherwise have been possible" is perhaps more aptly understood to mean "more than would otherwise have been possible, given that we are no longer willing to invest as much in the civilian space program as we used to be." Had competition as a primary motivator for space activities faded into obscurity, and would cooperation be the rule of the day for post-Cold War space policy makers? The remainder of the book will seek to answer these and related questions.

Finally, what does this era of "close work together" truly hold for man's future progress in space? Chapters Six, Seven and Eight will focus on post-Cold War developments which paralleled the improvement in U.S.-Russian relations. Examining some of the obstacles to cooperation in space reveals that while the improved international climate is a necessary condition, it is not a sufficient precondition for sustained, significant space cooperation. Moreover, the rise of commercial space interaction began to have greater implications for humans' future in space.

Notes

1. *Public Papers of the Presidents of the United States: Ronald Reagan, 1981* (Washington: GPO, 1982) 57. Hereafter referred to as *PP of RR*.

2. *Congress and the Nation* vol. 6, 201-16.

3. Reagan announced the initiation of the Star Wars program on March 23, 1983, describing the new anti-ballistic missile system as offering "a new hope for our children in the 21st century." *PP of RR, 1983*, book 1 (Washington: GPO, 1984) 437.

4. *Weekly Compilations of Presidential Documents* vol. 19, no. 10 (14 Mar. 1983) (Washington: GPO, 1983) 369. Reagan first used the term "evil empire" in a March 8, 1983 address.

5. Dulles' "rollback" or "liberalization" policy was aimed at freeing Central and Eastern European nations from the Soviet bloc. John Lewis Gaddis, *Strategies of Containment: A Critical Appraisal of Postwar American National Security Policy* (New York: Oxford UP, 1982) 155-56.

6. Murray Sale, "Closing the File on Flight 007" *The New Yorker* 13 Dec. 1993: 91. The airliner's black box was recovered by Soviet divers, and, under order of Andropov, then kept secret. Andropov knew by November 18, 1983 that the airliner was not on an espionage mission, but ordered that this information be kept secret. Finally, in January 1992, Russian President Boris Yeltsin ordered the flight's recordings handed over to the International Civil Aviation Organization (I.C.A.O.), the aviation arm of the United Nations. The I.C.A.O. report confirmed that the Korean airliner had accidentally flown off course. Soviet military commanders had at least temporarily mistaken the civilian aircraft as a standard American RC-135 reconnaissance mission. Despite seeing the plane's flashing navigation lights and being unable to confirm plane's type, the Soviet commanders ordered the plane shot down as per the recently enacted (November 24, 1982 and May 11, 1983), strict Soviet Air Codes. Sale, "Closing" 91-100.

7. *PP of RR, 1983*, book 2 (Washington: GPO, 1985) 1223-25.

8. Garthoff, *Détente* 1016-17.

9. Pages 200 and 201 (Appendix A, Charts 3 and 4) details these trends.

10. Soviet grain output was 237 million tons in 1978. Although varying weather and other conditions affected subsequent years' output, the general trend was decline and stagnation: grain output was 179, 189, 158 and 187 million tons in 1979, 1980, 1981 and 1982, respectively. Crop output grew by 6.6% from 1982 to 1983, only to drop again by 1.8% and 0.9% in 1984 and 1985, respectively. Meanwhile the USSR imported large quantities of grain: Soviet spending on grain imports had grown from $700 million in 1970 to $7.2 billion in 1980, and the Soviets purchased 31, 46 and 32 million tons abroad in 1979, 1981 and 1982, respectively. These purchases became more difficult and more expensive after President Carter imposed an embargo on the Soviet Union after the invasion of Afghanistan (Chapter Four, pages 87-88).

The problems caused by drops in production were exacerbated by inefficiencies in processing and transportation; as late as March 1989 Gorbachev estimated that 40% of agricultural produce was lost. In the late 1980s Soviet officials estimated that 42% of all collective and state farms were either operating at a loss or making a negligible profit and, as a result, required substantial subsidies. Marshall I. Goldman, *What Went Wrong with Perestroika* (New York: W. W. Norton, 1991) 63. Karl-Eugen Wädekin, "Agriculture," *Gorbachev and Perestroika*, ed. Martin McCauley (London: Macmillan, 1990) 82. Neil Felshman, *Gorbachev, Yeltsin and the Last Days of the Soviet Empire* (New York: St. Martin's Press, 1992) 102-4. Astrid von Borcke, "Gorbachev's Perestroika: Can the Soviet System be Reformed?" *Gorbachev's Agenda*, ed. Susan L. Clark (Boulder: Westview Press, 1989) 47-48. Stephen White, *Gorbachev and After* (Cambridge: Cambridge UP, 1992) 117-18.

11. The INF treaty eliminated nuclear missiles with ranges between 500 and 5500 kilometers. This affected 859 American missiles and 1836 Soviet missiles.

12. The START negotiations began in 1981 and stemmed from the intermediate-range nuclear forces talks of the late 1970s, but had been essentially suspended by the end of 1981. Garthoff, *Détente* 1022-23.

START 1 reduced the number of U.S. and Soviet warheads to 8500 and 6500, respectively; START 2 reduced these numbers to 3500 and 3000 and also eliminated land-based Multiple Independently targetable Reentry Vehicles (MIRVs).

13. Gorbachev, as allowed by constitutional changes, was elected President of the USSR on March 15, 1990 by the Congress of People's Deputies. The vote was 1329 for, 495 against. "Mikhail Sergeevich Gorbachev is Elected President of the USSR," *Vremya*, (television program) Moscow, 15 Mar. 1990.

14. *Public Papers of the Presidents of the United States: George Bush, 1992-93* (Washington: GPO, 1993) 2222-23.

15. One of the main reasons not all of the aid was delivered as promised is that the I.M.F., World Bank and European Bank for Reconstruction and Development concluded on several occasions that Russia was not meeting criteria they had set for aid delivery (e.g., cutting its inflation, budget deficit and subsidies). Thomas L. Friedman, "U.S. Asks Allies to Help Speed I.M.F. Aid to Russia," *New York Times* 1 Feb. 1994: A6. Steven Greenhouse, "Seven Offer Moscow Technical Help," *New York Times* 18 July 1991: A1. World Bank Policy Research Department, "Group of Seven Opens New Lending Windows for Russia," *Transition* Apr. 1993.

"Russia in Need," *The Economist* 15 Jan. 1994: 16.

16. "Agreement Between the United States of American and the Union of Soviet Socialist Republics Concerning Cooperation in the Exploration and Use of Outer Space for peaceful Purposes," 15 Apr. 1987. NHO 1.

17. Chapter Four, page 79, discusses the first edition.

18. *PP of RR, 1987*, book 1 (Washington: GPO, 1989) 379.

19. "Glasnost in Space," *Science News* 7 Sep. 1991: 156.

20. Brian Dunbar, "Successful U.S.-Russian Ozone-Monitoring Mission Appears Over," *News Release* 95-11, 2 Feb. 1995. NHO 6.

21. Craig Covault, "NASA, Soviets Discuss Joint Environmental Missions," *Aviation Week and Space Technology* 7 Oct. 1991: 68. Hereafter referred to as *Aviation Week*.

22. Stephanie Sansom, "Looking for Alien Life," *Nature* 22 Aug. 1991: 653. Leonard David, "Electrifying: Soviet Show and Tell," *Ad Astra* Dec. 1991: 5-6.

23. "Declaration by the Heads of State of the Republic of Belarus, the RSFSR and Ukraine" and "Agreement Establishing the Commonwealth of Independent States," *International Legal Materials* vol. 31, no. 1, Jan. 1992: 142-46.

24. "Protocol to the Agreement Establishing the Commonwealth of Independent States," *International Legal Materials* vol. 31, no. 1, Jan. 1992: 147. Georgia and the Baltic republics of Estonia, Latvia and Lithuania did not join.

25. David P. Hamilton, "U.S. and Russia Proceed Cautiously," *Science* 26 Jun. 1992: 1756. "Joint U.S.-Russian Missions Planned," *Facts on File* 27 Aug. 1992: 628. Energia is the scientific and industrial complex which produces Russian (and produced Soviet) spacecraft. Its post-Cold War responsibilities are further discussed in Chapter Eight, pages 159-62.

26. Leonard David, "U.S.-Russian Exchange Program," *Ad Astra* Sep./Oct. 1992: 6.

27. Daniel S. Goldin and Yuri N. Koptev, "Memorandum of Discussion: On Civil Space Cooperation," Moscow, Jul. 1992. NHO 8. The structure of the post-Soviet civilian space program is discussed in Chapter Six, pages 133-34.

28. Chapter Eight discusses these and other similar contracts in more detail.

29. The mission's timetable has limited flexibility, as it takes advantage of specific launch windows especially well-suited to a rendezvous with Mars.

30. William Gande, "Rovers Tested for International Mars Mission," *Ad Astra* Sep./Oct. 1992: 16. Louis Friedman, "Return to the Martian Surface," *Ad Astra* Sep./Oct. 1992: 29-32. Craig Covault, "Mobilize 'Space Year' to Save Russian Mars 94 Flight," *Aviation Week* 13 Jan. 1993: 35-36. Craig Covault, "Mars Strategy Begs for Direction," *Aviation Week* 5 Oct. 1992: 25-26. James R. Asker, "Mars 94 to Carry Novel U.S. Experiment," *Aviation Week* 12 Apr. 1993: 53. Craig Covault, "Mars Exploration Invites Global Space Cooperation," *Aviation Week* 12 Apr. 1993: 56-57. "Mars Rovers on the Move," *Sky and Telescope* May 1992: 487-88.

31. For example, Russia is responsible for launches and orbiters, France for the Mars Balloon, Germany for the orbital television system, Finland for instrumentation electronics and central processing units, Latvia for radar, Hungary for rover computers, Kazakhstan for the launch facility, Ukraine for tracking and the United States for the compact rover. Friedman, "Return" 30.

32. Chapter Six discusses the financial crisis of the late 1980s/early 1990s in the Soviet Union/Russia.

33. While discussions took place, the future of this project was anything but certain as of early 1994. President Bush's original goal of landing an American astronaut on Mars by 2019 had already been pushed back; cooperation with Russia could alter this delay. In a February 1992 appearance before the Senate Subcommittee on Space, Yuri Semenov, Energia's General Director, called for a joint Russian-U.S. human mission to Mars with the moon as a take-off point. "Russians Offer Use of Mir Space Station as a Prelude to Mars," *New York Times* 23 Feb. 1992: A26. As of early 1994, U.S. plans call for using space station Alpha as the take-off point for any manned mission to Mars.

34. Asker, "Mars" 53.

35. Apparently the problem was that NPO Lavochkin was having difficulty funding the launch vehicle's construction. Michael A. Dornheim, "Russians Press Mars 94, But Risks Still Run High," *Aviation Week* 3 Jan. 1994: 22-23. Peter B. de Selding, "Funding Needed to Ensure Mars Mission," *Space News* 17-23 Apr. 1995: 21.

36. Ben Iannotta, "Officials Strive to Keep Mars Together Alive," *Space News* 17-23 Apr. 1995: 20.

37. Peter B. de Selding, "Mars Mission Averting German Astronaut's Flight to Mir," *Space News* 27-5 Feb./Mar. 1995: 11.

38. Future U.S. Mars Surveyor missions were planned to include two launches in 2001, 2003 and 2005. Ben Iannotta, "JPL's Mars Pathfinder Program Taking Shape," *Space News* 19-25 Dec. 1994: 9. "NASA, Russia, Partners Juggling Mars Missions," *Space News* 17-23 Apr. 1995: 20.

39. Craig Covault, "U.S., Russia Plan New Mars Mission," *Aviation Week* 6 Jun. 1994: 24.

40. Roger Bourke, Personal Interview, 12 Apr. 1995. Dr. Bourke was the Director of the Jet Propulsion Laboratory's International Programs Integration Office.

41. Bruce A. Smith, "U.S.-Russian Flights to Planets Discussed," *Aviation Week* 20 Jun. 1994: 60.

42. Soviet and Russian probes were based on bulky military satellite buses. Robert Staehle, Personal Interview, 11 Apr. 1995. James Randolph, Personal Interview, 11 Apr. 1995. Drs. Staehle and Randolph were both JPL officials involved with the Fire and Ice program. See also: Aleksandr Pakhomov, "Russia to Help NASA Launch Probes to Pluto, Sun," Moscow *TASS* (11 May 1994) *FBIS* 11 May 1994: 19. Vladimir Rogachev, "Space Project With U.S. to Reach Sun, Pluto," Moscow *TASS* (11 Apr. 1994) *FBIS* 13 Apr. 1994: 30-31. "Pluto Flight Considered," *Aviation Week* 9 May 1994: 27. Leonard David, "Russia Considers Accompanying U.S. in Pluto Exploration," *Space News* 2-8 May 1994: 14.

43. Debra J. Rahn, "NASA and Russian Space Agency," *News Release* 94-101. 23 Jun. 1994. NHO 28.

44. James R. Asker, "Cosmonaut Upstages Shuttle Payloads," *Aviation Week* 31 Jan. 1994: 58-61. Warren E. Leary, "Shuttle Lifts Off on First Mission With a Russian," *New York Times* 4 Feb. 1994: A17.

45. There were to be two three-month visits and three six-month stays. William Harwood, "Tight Budget May Limit U.S. Trips to Mir," *Space News* 17-23 Jan. 1994: 19. NASA Office of Life and Microgravity Sciences and Applications, "U.S./Russian Cooperative Human Space Flight Program, Draft Schedule," 1 Nov. 1993: 24-26. NHO 25.

46. *PP of RR, 1984*, book 1 (Washington: GPO, 1986) 90.

47. *Congress and the Nation* vol. 8, 899.

48. NASA's budget pressure got even higher after the original decision to merge the American and Russian space stations: NASA's budget dropped to $13.7 billion in 1994, and was projected to drop from $14.2 billion in 1995 to $14.1, 14.0, $13.7, $13.5 and $13.3 billion in 1996, 1997, 1998, 1999 and 2000, respectively (moreover, none of these figures count inflation) *Budget of the United States Fiscal Year 1996* (Washington: GPO, 1995).

49. James R. Asker, "Space Station Key to NASA's Future," *Aviation Week* 15 Mar. 1993: 83.

50. Statement by the President, The White House, 17 Jun. 1993. NHO 3.

51. The U.S. record for time in space (through March 1994) was set at 84 days in 1974 when astronauts Carr, Gibson and Pogue returned from the Spacelab space station. In contrast, Soviet cosmonauts logged stays of 96, 139, 160, 175, 185, 211, 237, 313, 326 and 366 days in flights ending in 1978, 1978, 1987, 1979, 1980, 1982, 1984, 1992, 1987 and 1988, respectively. Moreover, the Russian Mir 2 designs and the U.S. Freedom plans closely resembled those of the highly successful Mir 1. Mir 1, which was launched in February 1986 and expanded in March 1987, November 1989 and May 1990, surpassed its expected lifetime, hosting fourteen piloted missions through early March 1994.   Wayne C. Thompson and Steven W. Guerrier, *Space: National Programs and International Cooperation* (Boulder: Westview Press, 1989) 133. National Aeronautics and Space Administration, *Aeronautics and Space Report of the President, Fiscal Year 1992 Activities* (Washington: National Aeronautics and Space Administration, 1993) 90, 93. Jeffrey M. Lenorovitz, "Proven Technology is Cornerstone of Mir 2," *Aviation Week* 23 Aug. 1993: 60-62.

52. Craig Covault, "Russia Forges Ahead on Mir 2," *Aviation Week* 15 Mar. 1993: 26-27.

53. Marcia Smith, telephone interview, 9 Mar. 1994.

54. *Congress and the Nation* vol. 8, 874-79 and 897-902.

55. Richard Nixon, "Clinton's Greatest Challenge," *New York Times* 5 Mar. 1993: A21. President Clinton and former President Nixon met to discuss aid to Russia on March 8. Thomas L. Friedman, "One Topic, Several Agendas and Clinton and Nixon Meet," *New York Times* 9 Mar. 1993: A1.

56. Elaine Sciolino, "Christopher Spells Out New Priorities," *New York Times* 5 Nov. 1993: A8. Warren Christopher was Secretary of State under Clinton. Michael R. Gordon, "Moscow Is Making Little Progress In Disposal of Chemical Weapons," *New York Times* 1 Dec. 1993: A1.

57. William J. Broad, "Daniel S. Goldin: Bold Remodeler of the Drifting Space Agency," *New York Times* 21 Dec. 1993: C8.

58. "Russian Space Program Seeks Needed Funding" *Aviation Week* 23 Aug. 1993: 24.

59. Leonard Nikishin, "Russia Has Given the Astro-Vehicle to America," *Obshchaya Gazeta* 8 Apr. 1994: 8. As translated by *JPRS*. "Alpha Station Said to Service Only U.S. Interests," *JPRS* 16 May 1994: 30.

60. Under the agreement regarding Russian entry into the commercial launch market, Russia was permitted to launch eight Western satellites to geostationary orbit through the year 2000 (launches of satellites to lower-earth orbit were to be considered separately on a case-by-case basis). Chapter Eight, pages 177-82.

61. The White House, Office of the Vice President, "Remarks by the Vice President in Signing Ceremony with Prime Minister Chernomyrdin of Russia," 2 Sep. 1993. NHO 9.

6

Obstacles to Cooperation in Space, Post-Cold War: A Nation in Disarray

Although the breakup of the Soviet Union and the end of the Cold War facilitated improved Russian-American space cooperation, the Soviet collapse also caused problems which threatened the new attempts at collaboration. From the late 1980s to the mid-1990s the Soviet/Russian economy fell deeper into a damaging cycle of deep recession and high inflation, and during the first half of the 1990s Russia struggled to attain political stability while the Russian space science and civilian space programs found themselves in the midst of financial crisis, brain drain and organizational disarray. This chapter shows that these severe problems were a part of the overall crisis of (former) Soviet science, and were exacerbated by the conservative Russian Academy of Sciences' resistance to reform. Together these difficulties threatened to undermine the cooperation facilitated by the post-Cold War improvement of U.S.-Russian relations.

Socioeconomic Crisis in the Former Soviet Union

The fall of the Soviet Union resulted largely from and served to worsen the nation's economic plight. The Soviet Union's gross national product shrank by four to five percent in 1990 and by twenty percent in 1991, and, when it collapsed, Russia inherited an $84 billion debt.[1] This debt made economic revival for Russia all the more daunting.

Russia's downward spiral continued through the mid-1990s. Official gross domestic product fell by 18% in 1992, 12% in 1993 and 15% in 1994; during the first half of 1995 the GDP fell by 4-6%. Russian industrial output fell some 14% in 1993 and 21% in 1994, leaving it at less than 45%

of 1991's output level--in other words, Russia's output more than halved in only three years. As the Russian government struggled to set its budget in early 1994, Prime Minister Chernomyrdin worried publicly that $35 billion of the $105 billion budget would come out of deficit spending, however the government managed to make it through November 1994 with a (still rather large) budget deficit of 10.7% of GDP.[2] Russia's foreign debt remained more than $80 billion by mid-March 1994.[3] In order to make ends meet Russia borrowed heavily. By June 1995 Russia's debt had ballooned to $130 billion.[4]

Russia's monthly inflation soared to 345% in January 1992 (following a sharp reduction in wholesale price controls) and averaged 2,500% for all of 1992.[5] According to official statistics, inflation averaged over 21% monthly from February 1992 to January 1994, totalling 840% in 1993.[6] The Director of the CIA, R. James Woolsey, believed that these official statistics severely understated inflation, which he estimated at 90% for January 1994 (versus the official statistic of 22%).[7] Even according to official statistics inflation remained high in 1994--220%. During the first five months of 1995 inflation rose 67%.[8]

Reflecting these poor conditions the ruble's value tumbled rapidly: at the end of April 1991, 1.6 rubles were worth one American dollar; in late March 1993 the official exchange rate was 668 rubles per dollar, and, in mid-January 1994 one dollar traded for 1,607 rubles.[9] On October 11, or "Black Tuesday," the ruble fell 27%. Black Tuesday outraged Russia's public and forced Yeltsin to fire his top economic policy makers (with the exception of Prime Minister Chernomyrdin); this purge worried Western observers, who feared that the next round of economic officials would be less reform-minded. Still, reform continued, and the ruble recovered partially from this dramatic drop in the ensuing weeks, finishing the year at 3,550 rubles to the dollar.[10]

Russia's economic recovery was further hindered by a large-scale flight of hard currency. The Russian government estimated in early 1994 that Russians had deposited $24 billion in foreign bank accounts in the previous two years.[11] This figure is comparable to the $38 billion in aid the West had sent Russia during that same period.[12]

At the same time efforts at privatization, the operation of private enterprises and the establishment of cooperative ventures with Western concerns were hampered by the rise of organized crime. Russia's Analytical Center for Social and Economic Policies estimated in early 1994 that more than three-quarters of the country's private enterprises were forced to pay 10-20% of their earnings to criminal gangs, and the official government statistics organization, Goskmostat, estimated that the "shadow economy" accounted for twenty percent of the gross domestic product. Companies often tempered this hardship by evading their taxes, thus exacerbating the nation's budget deficit. Many observers were encouraged when during the first half of 1995 the government managed to collect 36% of the period's taxes (compared with 20% during the same period in 1994)![13] The organized thugs hit Russia's newly evolving banking system especially hard. Many of the country's 1800 commercial banks were controlled by organized criminals, and those banks which resisted mafia influence paid heavily: eight leading bankers were assassinated between February 1993 and February 1994. Criminal threats often discouraged would-be Western partners: for instance, one American businessman arrived in Moscow seeking to explore a joint venture, only to be met at his hotel by five men wielding pistols and data about his company's earnings. After the racketeers demanded seven percent of his firm's future earnings, he simply returned to New York without conducting any business.[14]

Beneath all of these statistics were suffering Russians. Some Russian academics at the Academy of Sciences estimated that one in five Russians did not earn enough to buy basic foods in July 1995. These figures were similar to other estimates, which estimated that a staggering 35% of the population's income was below subsistence levels in 1992; these estimates put the same figure at 29% for 1993, 26% for 1994 and 29% for 1995.[15] A small portion of the population grew richer while the majority of Russians saw their standards of living drop significantly. The Duma's Economics Committee estimated that the poorest ten percent of the population lacked the food they needed for basic nutrition.[16] The average monthly income fell from $87 in November 1994 to $72 in March 1995. Meanwhile, although official unemployment rates hovered around three to four percent, actual unemployment was estimated to be between thirteen and seventeen percent in mid-1995.[17] The military-industrial complex was hit especially hard--roughly 2.5 million defense workers lost their jobs between 1990 and mid-1995.[18] Even those who worked were often not paid for months: estimates of workers who were not paid ran as high as twelve million in mid-1995. Social tension rose with all this hardship, boiling over increasingly frequently in strikes. There were, for example, more strikes during the first quarter of 1995 than in all of 1994.[19] Perhaps the most stark evidence of Russians' depression is the fact that the country's suicide rate grew dramatically to the world's third-highest, until almost one-third of all unnatural deaths were suicides.[20]

As reflected in demographic trends, this general economic malaise had a dramatic effect on the Russian population. Families, impoverished and/or fearing even worse economic times, had fewer children. Every year from 1987 to 1994 Russia's birth rate dropped while its death rate, after remaining fairly constant from 1986 to 1988, increased markedly from 1989 to 1994. By the end of 1994 Russian women lived an average of 71 years, while Russian men's life expectancy plunged from 64 in 1990 to a mere 57--the lowest rate of any country which reported to the World Health Organization. Demographers were stymied by these drops, since Russia was the first country in history to experience such sustained reversals in its health statistics, absent a major plague or war. Possible explanations included the decline of the health care system, the decline in the standard of living and the country's history of ecological abuse and the resulting increase in congenital diseases and birth defects. More than ten percent of all babies born in 1994 suffered from serious birth defects. Whatever the explanations, Russia's population began shrinking, dropping by 207,000 in 1991 and by nearly 800,000 to 148.4 million in 1993. With its economic base withering away, Russia's economic recovery remained an uphill battle.[21]

Still, there were signs of the economy bottoming out by the end of 1995. By June 1995 inflation had dropped to a monthly rate of 6.7%--still far higher than official goals of 1-2%, but significantly lower than most previous months.[22] Moreover, although industrial output fell roughly 21% in all of 1994, it had actually fallen approximately 24% in the first quarter of that year, recovering in the last three-fourths of the year. Prime Minister Chernomyrdin noted that during the first half of 1995 industrial output fell only three percent, and pointed to the relative stabilization of the ruble and the reduction of inflation as signs that the economy was improving. He also cited statistics saying that the gross domestic product actually rose 3% in June 1995 and that industrial output rose 2% during the same month.[23] These trends in output did continue, and output fell a total of only 3% in 1995; the GDP shrank only 4% that year. Perhaps most encouraging was the fact that housing construction actually grew by almost 10% in 1995.[24]

In early August 1995 the government approved the 1996 draft federal budget, which aimed to lower the budget deficit to 3.9% of GDP. The government hoped to cut its rate of borrowing from $12 billion in 1995 to $9 billion in 1996.[25] However, these encouraging signs were dimmed by other developments, such as estimates that salaries' real value dropped by one-third from March 1994 to March 1995, and predictions that 1995's grain harvest would be 70 million tons, down from 81.3 million tons in 1994 and 99 million tons in 1993.[26]

Political Stability Remains Elusive

The collapse of the Soviet Union was sudden and chaotic. Gorbachev had wanted to liberalize and decentralize the massive Soviet empire, but he was in fact a devout Communist who had every intention of holding together the Soviet Union. However, he had unleashed forces which were beyond control, and these forces grew rapidly as they gained inertia.

The reverberations of the collapse of the Soviet Union were felt around the world. The Soviet government had managed to keep up its facade of an advancing state remarkably well, considering the internal strife from which we now know it suffered. While government and academic experts in the West scrambled to adapt to the rapid changes, (former) Soviet politicians rushed to secure their own power bases. The result was political chaos. Power cascaded more than Gorbachev had ever hoped, away from the central government and down to republic and local officials. Yeltsin's government had to struggle to grab the reins of power, and once it did, it fought to give those reins some power.

Years later, after the dust had (more or less) settled, the former Soviet Union remained, by many standards, unstable. For a couple of years after the collapse of the USSR Russian government and law functioned (or failed to function) based on a largely ad hoc combination of Soviet and Russian law. The Communist Party of the Soviet Union, which had served as the organization and implementation backbone of the state, no longer existed. Even after he legitimized his presidency with popular support (such as a strong showing in a popular referendum held on April 25, 1993), international recognition and a reasonably effective government apparatus, Boris Yeltsin faced many of the same dilemmas as his Soviet predecessor: he was basically a reformer attacked from one side for not liberalizing enough, for being a dictator in sheep's clothing; and from the other side for abandoning Russia's great heritage and loosening up too much.

As the post-Soviet Russian government evolved, it was unclear how effectively the state could govern. In 1993 political battles polarized around the tradeoffs between high state spending on subsidies on one hand and fiscal prudence and price and output stabilization on the other. The parliament pushed for financial infusions from the state, while Yeltsin's government became increasingly intent upon cutting deficit spending. The clash was fierce and marked by a lack of willingness to compromise. The government and parliament each managed to approve a different budget nearly each month during the first nine months of 1993. Toward September their different budgets were further apart instead of closer together, and on September 21 Yeltsin made good upon earlier threats, suspending parliament.

However, the parliament would not be so easily dismissed. Russia's government was young and undefined, and, unlike other parliamentary governments with a strong executive and a long tradition of organized executive-legislative battles, it lacked the ability for smooth transitions from government to government. The parliament, led by (former) Vice

President Aleksandr Rutskoi and (former) Duma Speaker Ruslan Khasbulatov, rebelled against Yeltsin's edict, attempting instead to impeach the president and refusing to close parliament. The previous months of bitter fighting had hardly set a precedent for negotiation, and despite attempts at a peaceful resolution of the impasse, the executive-legislative standoff quickly broke into armed conflict. People in and out of Russia watched each development tensely, and they kept an especially close eye on the Russian military, which possessed the power to hand victory to either side.

On October 3 the country broke into something approaching civil war when some five to ten thousand Russians broke through a police cordon and surrounded the White House (where the parliament was holed up). Rutskoi and Khasbulatov thanked the people for their "victory," and exhorted "their people" to further rebellion: "I call on our glorious soldiers to seize tanks and take the Kremlin by storm!"[27] The crowd responded to Rutskoi's specific urging, taking over the Mayor's office and the Ostankino television station. At this point, observed the *Washington Post*,

> it seemed as though a Bolshevik-style revolution was unfolding. It appeared, that is, that a mere 5,000-10,000 people--determined, ruthless and facing only apathetic opposition--could grab control of a nuclear-armed giant with 150 million people.[28]

Yeltsin, in his memoirs, describes the situation at that point as "dismal." He and his closest allies had been pushing, unsuccessfully, for the army to come up with a few thousand crack troops to suppress the rebellion. Again, Yeltsin writes,

> the army, numbering two and a half million people, could not produce even a thousand soldiers; not even one regiment could be found to come to Moscow and defend the city....we were surprised: why is the army finding it so difficult to react? Why is it so poorly obeying orders? Because the army had been pulled into pieces and everyone was jerking on his part. How fortunate that no mad colonel had popped up with a squadron of bombers and flown over Moscow to defend his [military] friend Rutskoi.[29]

Finally, on October 4 the top of the military abandoned its stance of neutrality and attacked the parliament in the White House. The military met resistance, and even shelled the White House with tanks.

The image of the stately White House smoldering, charred and pitted with holes left by tank fire was a poignant symbol of the state of Russia's government. Although it had a proud and large land to rule, Russia's government was teetering on the edge of dissolution and chaos.

Khasbulatov, Rutskoi and the other losers were arrested, and Yeltsin held true to his promise to try to reconstruct the government, calling for elections in December of that same year. However, Yeltsin's popularity had suffered from the violence in Moscow as well as from the drop in the standard of living. The result was elections which seemed the harbinger of further political confrontation. In the December 12, 1993 parliamentary elections the Liberal Democrats, led by the nationalist Vladimir Zhirinovsky, shocked Yeltsin allies and the world when they won the largest plurality of the popular vote, 22.8% (the largest reformist party, Russia's Choice, came in second with 15.4% of the popular vote; the Communists came in third with 12.3%).[30] Zhirinovsky had waged an effective campaign, seizing on the volatile issues of crime and the collapse of the Russian empire, and unabashedly bragged of his ambitions to take over Russia and restore and enlarge Her empire.

This public rebuttal began a string of Yeltsin concessions to Russia's more conservative politicians. These concessions were highlighted by a reshuffling of Yeltsin's cabinet designed to slow down the move toward a market economy. On January 16, 1994 reformist (i.e., pro-market) First Deputy Prime Minister and Economics Minister Yegor T. Gaidar resigned from President Yeltsin's cabinet. Reformist Finance Minister Boris Fyodorov followed on January 26, 1994.[31] Their departure increased the power of Prime Minister Viktor Chernomyrdin and the chairman of the Russian Central Bank, Viktor Gerashchenko, both of whom advocated increased spending despite the runaway inflation.[32] Then, in February 1994, in a sharp rebuke to President Yeltsin, the Russian Parliament granted amnesty to the leaders accused of leading an armed insurrection against the Russian government in October 1993.[33]

Despite the turmoil of 1993 and early 1994, the Russian government churned on for the rest of 1994, almost acquiring, despite all the bitter political fighting, a sense of normalcy. However, as the year progressed another hot spot boiled over: the secessionist Russian region Chechnya. Chechnya is a small region in southern Russia near Azerbaijan with a history of centuries of hostilities with the Russians. For months the Chechen leader Dzhokar Dudayev had flaunted the fact that Chechnya had declared full independence from Russia in 1991, and Yeltsin and his government staunchly refused to acknowledge the region's sovereignty. Dudayev and his forces declared independence, and on December 11, 1994 Yeltsin ordered some 40,000 Russian troops to invade Chechnya and quell the rebellion. Although the military operation was intended to take a few days, a combination of poor military readiness, planning and weather, combined with a severe underestimation of the Chechens' ability to resist, led to a military debacle for Russia. Weeks of fighting turned into months, and the Russians suffered embarrassing military setbacks.

Would-be Western allies of Russia were compelled by stories of heavy civilian casualties and other concerns to chide Russia publicly, and President Clinton's visit to Moscow was threatened. Clinton did attend the summit with Yeltsin, but he publicly called for a permanent ceasefire. Yeltsin responded that there were no "combat operations," going on in Chechnya--a misstatement at best. Later Yeltsin publicly reprimanded his military and security leaders for disobeying and misinforming him. It was not until late July 1995 (after a Chechen incursion into Russia itself) that the fighting had subsided enough for the two sides to sign a partial treaty, and small-scale fighting dragged on into August 1995. As of August 1995 it remained to be seen if some long-term resolution would be found.

The disaster in Chechnya worsened Yeltsin's already declining popularity. Russian politicians of all leanings criticized their president's actions. Meanwhile politics began to crystalize around the upcoming elections for parliament (December 1995) and president (June 1996). Although it was unclear if Yeltsin would seek reelection, there were increasing signs that his tenure was near an end. First, polls showed that only 2-3% of the population would vote for Yeltsin for president. Second, Yeltsin's health deteriorated markedly. On July 10 Yeltsin had a heart attack serious enough to hospitalize him for two weeks. However, in a show of trickery reminiscent of the days of Stalin or Mao Tse Tung, government press officials denied reports of the heart attack for several days, even releasing what seemed to have been a doctored and/or mis-dated photograph. Eventually aides admitted to the President's heart attack and publicized Yeltsin's condition.[34]

The Foundation for Space

Soviet space policy had always been partly a product of domestic influences. Instability in the early 1990s threatened to severely debilitate Russia's ability to support its space program. Achievements in space were the luxuries of an advanced state which was wealthy enough to push back the frontier of science, and Russia's wealth was by no means secure. The social, economic and political base upon which the Soviet Union built its space program continued to wither away at an alarming rate in the mid-1990s, and the space program struggled to avoid falling with it. As of mid-1995, it remained to be seen how long the Russians space program could remain relatively healthy, just at it remained to be seen whether or not Russia would begin to recover socially and economically and to establish a stable body politic. If Russia could not manage to feed and provide basic care for its people on earth, it would eventually lose its ability to put people and objects into space.

So how did Russia's space program weather these stormy times? Soviet and Russian spending on space plummeted during the late 1980s and early 1990s. According to the Stockholm International Peace Research Institute (SIPRI), overall Soviet space expenditures totalled 6.9 billion rubles in 1989, fell to 6.3 billion rubles for 1990 and remained at this level in 1991.[35] This leveling did not account for 1991's 250% annual inflation. The Space Research Institute's 1991 funding was increased by 70% to 250 million rubles, but, once again, this fell far short of the 450 million rubles which would have kept pace with inflation. Although hard figures were unavailable in mid-1995, there is reason to believe that space spending stabilized, at least as a percentage of gross national product, by 1993.[36]

These grim figures do not, however, paint a complete picture of the health of the post-Soviet space program in mid-1990s. Although space spending declined significantly, the actual meaning of this drop was attenuated by the fact that space industry workers' salaries rose more slowly than other industry workers, and by the fact that the Soviet Union had overproduced space hardware, thus allowing post-Soviet officials to draw upon their stock during the tough times.[37] In addition, much of the reduction in space spending in the early 1990s was due to the cancellation of the Soviet space shuttle Buran program. Finally, many of the Newly Independent States' space enterprises supplanted their government support with international commercial cooperation (Chapter Eight).

In fact, statistics which were available in mid-1995 suggested that Russia's space budget had stabilized to reasonable levels by the middle of the 1990s. In 1994 Russia spent approximately 0.05% of its gross national product on piloted space activities, which is comparable to the corresponding figure of 0.09% for the United States, especially if the Russian figure is increased by including foreign payments of hard currency for access to Mir.[38] Finally, in many senses the bottom line is launch rate, and Russia remained, through 1994, the world's busiest launch nation, launching 48 satellites successfully in 1994 (versus 26 by the U.S.) and 17 satellites in the first seven months of 1995 (versus 16 in the U.S.). Although this was a significant drop from previous years, this drop was somewhat ameliorated by Russian advances in satellite technology (e.g., the Conclusion's comments on pages 189-93 about changed reconnaissance satellite needs).

Russian Science: Sacrifice at the Economic Altar

Russia's future in space is related to the overall future of Russian science. Russia's space program draws upon the country's scientific and engineering talents and infrastructure in a variety of fields, and if Russian science suffered, the space program suffered with it.

Russian science was hit hard by the general economic collapse. The country was struggling to feed its people, provide social services and maintain its armed forces; science was, in comparison, expendable. As Roald Sagdeev, the Director of the Soviet Institute of Space Research (IKI) from 1973 to 1988, described his field in late 1991, "With the economy collapsing, there has to be a sacrifice at the altar."[39] Igor A. Nikolaev, the head of Science and Technology Policy at the Russian Ministry of Science, Higher Education and Technical Policy estimated that government funding for science comprised six percent of the Soviet Union's 1990 GNP, but only 1.9% of Russia's 1992 (shrunken) GNP.[40] Nikolaev's estimate was consistent with other statistics, which reported that in 1995 spending on scientific research and development was approximately one-eighth of the level in 1985. Between 1990 and late 1993 alone Russia's science budget shrank by roughly two-thirds in real terms.[41]

Russian military research and development budget was cut by 50% in 1991 and again by 80% in 1992; civilian research was hit even harder as government officials tried to preserve only the essentials of Russia's scientific community.[42] While Russia's science budget shrank by two-thirds from 1990 to late 1993, its scientific work force was reduced by only thirty percent. As a result, a greater and greater percentage of the scientific budget was spent on salaries while less and less was available for equipment and other necessities. Some estimated in 1995 that sixty percent of all money for research and development was spent on salaries.[43] In January 1992 the Russian Academy of Sciences canceled its support of subscriptions to foreign journals, severing this vital link to the world's scientific community.[44]

Astronomers were especially hard hit. Observatories, their budgets slashed, struggled to perform any real science. Some observatories could not cover basic utility bills, let alone pay for ongoing or new research programs. For example, Victor Abalakin, director of the renowned Pulkovo Observatory in St. Petersburg, complained that his electric bill for all of 1991 was 78 million rubles, while the same bill for the first ten days of 1992 was 118 million rubles. At this rate, Abalakin said, electricity would consume his entire budget.[45] Zhores Alferov, director of the Ioffe-Physico-Technical Institute (also in St. Petersburg) and a Vice President of the Russian Academy of Sciences confirmed that "many" institutions could not pay for utilities, photocopying, reagents and equipment maintenance and noted that many were forced to stop paying salaries in December 1991.[46] Academy salaries continued to be delayed in the summer of 1993.[47] As Viktor N. Sokol'sky, long-time historian of the Soviet space program described the situation, "Of course we're all glad the Cold War is over. Still, it's been very bad for us who work in space. Before, everything was planned, our budgets were certain....Now nothing is certain."[48]

As the country's financial ruin worsened and budgetary pressures increased, scientists saw their salaries' real values decline--that is, when their salaries did not go unpaid. This development marked a heightened level of Russia's scientific crisis. In 1991 there were over ten million Soviet citizens, counting support personnel and family members, dependent on the space industry for their livelihood. Ninety-five percent of these ten million were in Russia.[49] With their incomes dwindling many of these scientific workers were forced to scramble for essentials, and many became angry. On January 19, 1992 science workers in *Zvezdnyj Gorod*, the model Star City, staged a symbolic strike, protesting their five hundred rubles/month salaries.[50]

Indeed, across Russia scientists and scientific workers saw their salaries' value decline below that of less educated workers.' Although quantifying the salary differences is not an exact process, the following examples are illustrative: in January 1992 the average Russian Academy of Sciences salary was 1,000-2,000 rubles/month, compared to 2,000 rubles for a Moscow bus driver. Later estimates (April and September 1992, respectively) placed researchers' salaries between 1,500 and 3,000 rubles/month, compared to the minimum monthly wage for a skilled worker of 7,000 and 10,000 rubles.[51] In October 1992 engineers at the Scientific-Industrial Union *Molniya* (lightning) made the equivalent of ten dollars a month, or roughly one-fifth the salary of a Russian bus driver.[52] In 1993 the average scientists' wage was 38% lower than the average industry salary.[53] Scientists and engineers in the former Soviet Union were facing grim times.

The Russian Brain Drain

As scientists and scientific workers found themselves less and less able to provide for themselves and their families by plying their trades, they took advantage of their new freedoms and sought work elsewhere. As conditions worsened in the early 1990s, a brain drain resulted. The first aspect of the brain drain was internal and is hard to quantify. Scientists and technical workers, seeking higher salaries in other fields, abandoned science for other work in the former Soviet Union. Still other scientists maintained their offices and titles even as they spent the majority of their time on private, more profitable ventures. For example, in 1993 I met with a prominent Russian astronomer who had published scores of articles in all of the world's best astrophysical journals. He arrived to our meeting late and panting, his hands covered with oil and dirt. He apologized profusely: he had just gotten off work fixing automobiles.

The second type of brain drain was external. Tempted by opportunities abroad, former Soviet scientists left their *rodina*, or homeland. This emigration threatened to permanently damage the former Soviet republics' scientific capabilities. Igor Makarov, the last chief secretary of the Soviet Academy of Sciences, estimated that 564 research fellows of academic institutions left the Soviet Union in the first eleven months of 1991--more than twice the number that left in all of 1989. Russian Academy of Sciences President Yuri Osipov stated that at least five hundred Academy workers left for the West between December 1991 and April 1992.[54] In late 1993 the U.S. State Department estimated that 2,000, 2,500 and 3,000 Russian scientists and engineers emigrated to the United States in 1990, 1991 and 1992, respectively, and estimated that a still larger number would immigrate by the end of 1993.[55] By late 1991 Israel alone had accepted over 6,000 basic researchers from the former Soviet Union.[56] The Paris-based Organization for Economic Cooperation and Development (OECD) estimated that a total of just under 30,000 Russian scientists had emigrated by the autumn of 1993, a number consistent with other estimates.[57] It is likely that these numbers understate the reality of the situation, since not all scientists working abroad had officially emigrated; some, for example, were on indefinite or long-term leaves.

In 1989 the Soviet Union had 1.5 million scientists and scientific workers. This made the USSR host to one-fourth of the world's scientists and researchers.[58] By late 1993 Russia's scientific work force had, according to OECD estimates, dwindled to just over 600,000.[59] Given this large work force, the former Soviet Union could lose some scientists while perhaps remaining a first-rate scientific nation.

However, the effects of the intellectual exodus were exacerbated by the fact that the majority of the scholars emigrating were those who could compete with Western scientists for faculty and research positions. This competition was especially harsh due to falling post-Cold War scientific budgets in the United States--for example, 813 physicists applied for a single tenured opening at Amherst College in 1992.[60] As a result (former) Soviet scientists who succeeded in finding work in the West were often the best. The results of this selection process were potentially staggering: one estimate stated that as many as half of Russia's most-cited scientists had already left by the end of 1991.[61]

The brain drain affected many areas of Russian science. For example, forty researchers (thirty-six of them senior researchers) of the Moscow Institute of Molecular Genetics' two hundred scientists left Russia by April 1992.[62] Other examples include the Lebedev Physical Institute's theoretical division, where, of fifty-five scientists, four to five emigrated permanently and five to seven others left by May 1992, their returns uncertain,[63] the Moscow Kurchatov Institute of Atomic Energy, where twenty of fifty-four theoretical plasma scientists left for the West by June 1992,[64] the Institute of Cytology in Saint Petersburg where fifty of the institute's two hundred fifty scientists left the country by February 1993,[65] the Engelhart Institute of Molecular Biology in Moscow, where forty-nine of three hundred scientists left had been abroad for at least a year as of February 1993 (again, their returns were uncertain),[66] and the Landau Institute of Physics, which was, by April 1992, "half-empty."[67]

The brain drain hit Russia's space industry especially hard. The Russian Space Agency's General Director, Yuri Koptev, estimated that Russia's space industry lost 30% of its highly trained specialists and that 34% of Russia's top space scientists left Russian institutes by August 1993.[68] In 1992 the Russian missile and space industry lost roughly 80,000 employees, including ten percent of its production work force and thirty percent of its research staff.[69] Although it may well take years to gauge the effects of this massive mental migration, it bodes poorly for the future of Russia's space program.

Despite such foreboding, the post-Soviet brain drain was a predictable and probably even a healthy, if painful, process. As host to one-fourth of the world's scientists and engineers, the Soviet Union was a bloated scientific power. It is questionable whether any state, let alone an unstable one, could support such a heavy scientific and engineering population. The majority of the brain drain was in the military and industrial sectors, and it is these very sectors which most needed to downsize in the post-Cold War, post-Soviet environment. As the 1990s unfolded, there were signs that Russia would remain a scientific power, one with a more reasonably sized scientific work force. At the end of 1993, for example, Russia's research and development workers per 10,000 people was comparable to the levels in other major scientific powers:

Country	Scientists per 10,000 People
Japan	92
Russia	91
U.S.A.	76
Germany	59[70]

Organizational Problems: The Russian Academy of Sciences

In the early 1990s the future of space cooperation was also threatened by lingering problems with the organization of Russia's research and development and space infrastructure.

As the Soviet Union dissolved, so too did the Soviet Academy of Sciences. After heated debate, those who hoped for the maintenance of some form of inter-republic Academy gave up, and Russia assumed control of the Soviet Academy's Russian holdings and personnel. The plan to create a Russian Academy of Sciences (RAN) was begun on March 25, 1991, and the Russian Academy finally absorbed those parts of the Soviet Academy which existed on Russian territory on December 6, 1991.[71] The new RAN gained control of the over three hundred institutes which conducted the bulk of Russia's basic research.

Even as the Russian Academy inherited most of the Soviet Academy of Sciences' personnel and assets, so too did it acquire the Soviet Academy's tradition of resistance to change. The USSR Academy of Sciences was a lingering bastion of conservatism. It was one of only two learned societies which did not condemn the anti-Gorbachev August 1991 attempted coup as unconstitutional.[72] As Professor Loren Graham, an expert on Soviet and Russian science, writes,

> [the Soviet Academy had] cooperated fully with the secret police in controlling the travel abroad of its scientists who wished to attend scientific meetings, it was dominated by senior scientists who often ignored the needs of younger colleagues, it condoned the anti-Semitism that infected some of its research institutes, and it suffered from the bureaucratic rigidity endemic to all Soviet organizations.[73]

The members of the Academy long enjoyed privileged treatment and funding under the Soviet system, and, after the switch to the Russian system, they were slow to give it up. The Russian Academy inherited a tradition of cronyism. Instead of administering funding on the basis of merit, senior members routinely received priority treatment. This generational nepotism tended to stifle innovation and discourage younger scientists.

However, as the Soviet Union was collapsing, the Academy's established methods came into question. From December 10 to December 12, 1991, a "Conference of Scientists of Academic Institutions" was held in Moscow, largely to determine the future organization of the Russian Academy of Sciences. Some conference participants hoped to radically change the way Russia conducted science, including elevating the role of universities, which had historically played a minor role in basic research. However, their success was limited at best: although the "mild reformers," who sought to introduce elements of democracy into the Academy's existing administration won,

> as time goes on the new Russian Academy of Sciences (RAN) looks more and more like the old Soviet Academy of Sciences. Truly meaningful reforms did not take place....In the end, the system of privilege, perquisites, and authority that the old Academy represented turned out to be too powerful to permit genuine reform.[74]

In a further blow to would-be reformers the academy elected Yuri Osipov, a conservative "representative of the old military-industrial complex and a defender of the system of a dominant central academy" as its president.[75]

Despite this largely successful resistance to change, the battle to reorganize Russian science continued after December 1991. Boris Saltykov, a reformer, became Yeltsin's Minister of Higher Education and Technology Policy. In April 1992 Saltykov led the formation of the Russian Foundation for Basic Research (FBR). The FBR was modeled after the U.S. National Science Foundation and was intended to award scientists grants on the basis of merit. In early 1993 the new foundation distributed its first nine thousands grants. Although modest in scope, this was Russia's first multidisciplinary, Western-style grant agency.

Still, the foundation's success was marred by political battles between Osipov and RAN and Saltykov, Vorontsov and other reformists.[76] Calls for reform in Russia's scientific infrastructure grew more and more muted as the early 1990s progressed. Opponents of Saltykov and his proposed changes openly called for his removal, and, in late 1993/early 1994 it was rumored that he was in fact going to be fired. The shakeup came, but Saltykov stayed. Nonetheless, the apparent near-success of this attack on one of Russia's last great hopes for scientific reform reflected the fact that by March 1994, the "time when the academy seemed to be reforming had ebbed...[and] the establishment seemed to be prevailing."[77]

This failure to reform significantly threatened to greatly diminish Russia's scientific capabilities, and, as a long-term result, its overall capacity as a space power. Even as Russia's general economy struggled to adapt to the rule of the market, Russian science continued in its tradition of conservatism. The Academy of Sciences maintained its old system of seniority and patronage, and fiercely resisted proposals that it greatly diminish its work force. However, the fact remained that its ranks had become extremely bloated, and because its budget was dominated by salaries, the academy's ability to fund innovative science was significantly hampered.[78] It is ironic that, in its unwillingness to act upon the fact that Russia could no longer afford to fund the number of scientists it once did, the academy may well have been exacerbating Russia's scientific woes. Still, this irony is but a part of the larger economic struggle; as one Russian scientist commented on the hesitation to fire less productive workers, "We don't have a well-developed system of help for the unemployed."[79]

As of the mid-1990s Russia's general economic, social and political health, combined with the country's scientific community's budgetary woes, brain drain and continuing conservatism, posed potentially great problems for cooperation in space. If Russia's ability to sponsor its space program continued to diminish to the point where its efforts in space had to be significantly scaled back, it would have less to offer the United States in terms of cooperation.

Western Aid to Russian Science

The West responded to the crisis in Russian science, trying to prevent nuclear proliferation and pursue profitable ventures. The United States and Germany, afraid that top former Soviet nuclear scientists would be courted by nations trying to develop their own nuclear weapons, developed the Baker-Genscher initiative. As a part of this initiative, the United States agreed in February 1992 to provide $35 million, and Europe $25 million, to create "clearinghouses" for the former Soviet Union's top two thousand nuclear weapons scientists to do non-weapons related research.[80]   As CIA Director James Woolsey explained Western motivations in July 1993,

> delays in pay, deteriorating working conditions, and uncertain futures are apparently spurring Russian specialists to seek emigration despite official restrictions on such travel....[other nations such as China, India, Iraq, North Korea and Pakistan have attempted to recruit such scientists]...."[T]he combination of declining morale in the military, increased organized crime, and efforts by states like Iran seeking to purchase nuclear material or expertise will make these matters a major concern...through this decade and beyond."[81]

This program evolved into the International Science and Technology Center (ISTC), which opened in March 1994. Between March 1994 and the middle of 1995 the ISTC had spent nearly $70 million to support some 9,500 Russian scientists.[82]

In addition, American efforts, philanthropic, governmental and otherwise, attempted to aid Russian science and take advantage of Russia's scientific experience and abilities. In March 1992 the U.S. Department of Energy signed a contract with the Kurchatov Institute of Atomic Energy, agreeing to fund 116 scientists for a year of fusion studies. The total cost of the scientists' salaries and equipment is ninety thousand dollars--a fraction of the millions such an endeavor would cost in the United States.[83] Although ninety thousand dollars is a pittance to the Department of Energy, it accounted for nearly a third of Kurchatov's 1992 fusion budget.[84]

Philanthropic assistance took a variety of forms. Through April 1995, the National Science Foundation gave $5 million to Soviet scientists,[85] the American Physical Society and American Astronomical Society sent journals, hundreds of small grants and even computer equipment,[86] the

Stanford Linear Accelerator donated a $10,000 Department of Energy Grant to fund twenty theoretical physicists at the Physical Technico Institute in St. Petersburg,[87] the American Physical Society (with the help of the Sloan Foundation, National Science Foundation and George Soros) donated $700,000 directly to needy scientists,[88] the Howard Hughes Medical Institute gave $14 million dollars to support five-year research programs,[89] the Department of Energy signed a $10 million contract with the Joint Institute for Nuclear Research in Dubna, Russia to develop supercollider technology[90] and the MacArthur Foundation announced it would donate $3 million to Russia; part of the MacArthur money was dedicated towards aiding Russian environmental research.[91] In the biggest single effort American entrepreneur George Soros dedicated $112.5 million to the "International Science Foundation for the former Soviet Union." As of April 1995 Soros' foundation had disbursed approximately $90 million of its funds, and Mr. Soros was considering allocating additional aid for former Soviet science if other sources would match his contributions.[92] Although this assistance may prove to be a band-aid for a gaping wound which can only be healed by its victim, it constituted a significant expansion of cooperative efforts.

Organizational Problems: The Russian Space Program

Organizational problems of a different nature plagued the former Soviet space program. The Soviet space program was entirely military, and existed, along with all other missiles programs, within the Ministry of General Machine Building. The dissolution of the Soviet Union and the privatization of state-owned enterprises left various Russian organizations vying for control of the Russian space program. In the past the military simply told enterprises what to produce; in the new environment, enterprises had to bid for contracts with military, civilian and foreign customers.

After the Soviet Union dissolved Russia's civilian space program became officially the dominion of the governmental Russian Space Agency (RSA). The RSA's General Director, Yuri Koptev, worked to expand his agency's authority, but his goals clashed with those of other space enterprises. Relations became especially strained between the Russian Space Agency and Energia, Russia's largest space enterprise (for a detailed description of Energia, see Chapter Eight, pages 159-62). For example, under governmental contract Energia employed all civilian cosmonauts (military cosmonauts continued to work for the Ministry of Defense) and the staff of the piloted flights area of the Flight Control Center. However, the Flight Control Center reported to the Russian Space Agency. Overlapping and interdependent responsibilities such as these exacerbated turf battles. In other conflicts, the Khrunichev, Glavkosmos and KB (Engineering Bureau) Salyut enterprises fought over control of the Proton launch vehicle.[93] The highly centralized, top-down Soviet system had been replaced by a vaguely defined collection of enterprises.

Unlike NASA, which employs tens of thousands of people, RSA employs a few hundred people. RSA was established as a "general contractor," and is basically a shell organization intended to implement Russian civil space policy through the management of contracts.[94] In August 1993 Russia adopted the Law on Space Activities in the Russian Federation, which declared that space activity in Russia would be:

- dominated by the practical needs of society;
- performed on a competitive basis;
- open to the public;
- subjected to independent review.

According to this same law, Russian military space activity was intended to:

- provide support to ground operations;
- monitor compliance with international agreements;
- provide warning of attack;
- deter potential aggression.[95]

The role of the executive and legislative branches of the Russian government in setting space policy were also unclear in the early 1990s. The Russian Space Agency did not even place the executive branch's Ministry of Science, Higher Education and Technology Policy on its 1992 organizational chart describing the structure of the Russian space program, while the science Ministry claimed that the Ministry of Finance required its approval for the RSA's budgets. In 1991 and 1992 Russia's Supreme Soviet (through its Commission on Transport, Communication, Informatics and Space) tried to assert greater control over the space budget, succeeding in getting civilian space expenditures identified as a specific line in the Russian budget for the first time.[96] The Russian Duma created by elections on December 12, 1993 formed a Committee on Science and Technology: Nikolai Vorontsov chaired this committee, but, as of early 1994 its role in setting space policy remained unclear.

In addition, the former Soviet space program was more than Russian. Chapter Seven examines how the breakup of the Soviet Union into the Newly Independent States posed challenges for the future of U.S.-NIS cooperation in space.

Notes

1. U.S. Congress, Joint Economic Committee, *Global Economic And Technical Change* Hearing 16 May and 28 Jun. 1991, S. Hrg. 102-586, pt. 1 (Washington: GPO, 1991) 152. U.S. Congress, Joint Economic Committee, *Global Economic And Technical Change: Former Soviet Union and Eastern Europe, China* Hearing 8 Jun. and 27 Jul. 1992, S. Hrg. 102-586, pt. 2 (Washington: GPO, 1992) 94. Richard Nixon, "Clinton's Greatest Challenge," *New York Times* 5 Mar. 1993: A21.

2. John Thornhill, "Russian Output Halved in Three Years," *Financial Times* 31 Dec. 1994: 2. Anders Aslund, "Ruble Awakening," *Washington Post* 23 Apr. 1995: C4. Thomas Sigel, "Economic Decline in 1994," *OMRI Daily Digest* no. 19, pt. 1, 26 Jan. 1995. Celestine Bohlen, "Russian Economy Has I.M.F. Checkup," *New York Times* 18 Mar. 1994: A9. "GDP Down But Some Sectors Show Growth," *OMRI Daily Digest* no. 135, pt. 1, 13 Jul. 1995. "Chernomyrdin Describes Economy," *OMRI Daily Digest* no. 140, pt. 1, 20 Jul. 1995.

3. Michael Specter, "Russia Promises Budget Curb to Win a Loan of $1.5 Billion," *New York Times* 23 Mar. 1994: A1.

4. Michael Mihalka, "Davydov on Foreign Debts, Investments," *OMRI Daily Digest* no. 123, pt. 1, 26 Jun. 1995.

5. Steven Erlanger, "Inflation and Unpaid Bills Haunt Russia as it Works on '94 Budget," *New York Times* 14 Feb. 1994: A9. Aslund, "Awakening."

6. *"V Zerkale Statistiki,"* ("Statistics in the Mirror") *Ekonimika i Zhizn'* (*Economy and Life*) Dec. 1992, no. 51: 1. Janet Guttsman, "Russian Economy Seen Contracting," *Reuter European Business Report* 24 Oct. 1994.

7. John Lancaster, "Russia's Economic Prospects Fade, CIA Director Tells Hill," *New York Times* 26 Jan. 1994: A2.

8. "Decline in Russian GDP Widened," *Wall Street Journal* 18 Jan. 1995: A10. "Retail Prices Increase in May," *OMRI Daily Digest* no. 100, pt. 1, 7 Jun. 1995.

9. James R. Asker, "Soviet Space Programs Bid for Funding in Weak Economy," *Aviation Week* 6 May 1991: 23. Suzanne Possehl, "Its Budget Slashed, Russian Seed Bank Fights for Its Life," *New York Times* 23 Mar. 1993: C4. Steven Erlanger, "Ruble Sinks Further in a Whirl of Political Rumors," *New York Times* 20 Jan. 1994: A3. Claudia Rosett, "Ruble's Slide Spotlights Power of Old-Style Central Bank Chief," *Wall Street Journal* 20 Jan. 1994: 1. February and March 1994 saw some degree of economic stabilization--inflation was less than 10% in February and was 9.9% in March, and 1,750 rubles traded for one dollar on March 23, 1994. It was unclear whether or not this trend would continue. Departed Finance Minister Fyodorov (pages 123) claimed that these months reflected reforms that had been instituted several months earlier and had since been abandoned. Thus, he predicted that inflation would rise and the ruble's rate would fall in months to come. Celestine Bohlen, "Moscow Rumors, True or Not, Show The Weakening of Yeltsin's Position," *New York Times* 27 Mar. 1994: A12. Specter, "Promises" A1. Boris Fyodorov, "Moscow Without Mirrors" *New York Times* 1 Apr. 1994: A27.

10. Thornhill, "Output."

11. Serge Schmemann, "Russia Lurches Into Reform, But Old Ways Are Tenacious," *New York Times* 20 Feb. 1994: A1.

12. Chapter Five, pages 97-98.

13. "Shadow Economy," *OMRI Daily Digest* no. 104, pt. 1, 30 May 1995. "But Tax Collections Improve," *Monitor* vol. 1, no. 42, 29 Jun. 1995.

14. "The High Price of Freeing Markets," *The Economist* 19 Feb. 1994: 57. Schmemann, "Lurches" A1.

15. "Twenty Percent of Russians," *Monitor* vol. 1, no. 63, 31 Jul. 1995. "Life During Reform: A Portrait in Numbers," *New York Times* 16 Jun. 1996: A6.

16. "One Russian in Ten," *Monitor* vol. 1, no. 28, 28 Jun. 1995.

17. "Average Russian Income Drops," *OMRI Daily Digest* no. 95, pt. 1, 17 May 1995. "Yeltsin Aide: Unemployment to Twenty Percent by Year's End," *Monitor* vol. 1, no. 39, 26 Jun. 1995.

18. "Military Industrial Complex in Trouble," *Monitor* vol. 1, no. 47, 10 Jul. 1995.

19. "Twelve Million Russians," *Monitor* vol. 1, no. 54, 18 Jul. 1994. "More Russians Out on Strikes," *Monitor* vol. 1, no. 32, 14 Jun. 1995.

20. Penny Morvant, "Russia Had Third Highest Suicide Rate," *OMRI Daily Digest* no. 100, pt. 1, 7 Jun. 1995.

21. Michael Specter, "Climb in Russia's Death Rate Sets Off Population Implosion," *New York Times* 6 Mar. 1994: A1. Schmemann, "Lurches" A1. Michael Specter, "Russians Dying Younger and Younger," *New York Times* 2 Aug. 1995: A1.

22. "Russia Leads CIS In Inflation," *Monitor* vol. 1, no. 63, 31 Jul. 1995.

23. "Output Falling More Slowly," *Monitor* vol. 1, no. 49, 11 Jul. 1995. "Chernomyrdin Describes."

24. "Life During Reform" A6.

25. "Government Approves 1996 Draft Budget," *OMRI Daily Digest* no. 151, pt. 1, 4 Aug. 1995.

26. Penny Morvant, "Deterioration in Social Situation," *OMRI Daily Digest* no. 92, pt. 1, 12 May 1995. "Russia Will Likely Import More Grain This Year," *Monitor* vol. 1, no. 51, 13 Jul. 1995. Russia's production of meat and milk fell by almost 25% during the first five months of 1995, and de-ruralization threatened Russia's long-term agricultural production. In 1960 there were 73,000 villages with fewer than 1,000 residents, but by mid-1995 there were fewer than 24,000, and 80% of the remaining occupants were over 35. Thomas Sigel, "Meat and Milk Production Decline," *OMRI Daily Digest* no. 123, pt. 1, 26 Jun. 1995. Michael Specter, "New Freedoms Churn a Quiet Russian Village," *New York Times* 14 Aug. 1995: A1.

27. "Mass Misjudgment," *The Economist* 9 Oct. 1993: 57.

28. Margaret Shapiro and Fred Hiatt, "Troops, Yeltsin Foes in Pitched Fight Outside Parliament," 4 Oct. 1993: A1.

29. Boris Yeltsin, *The Struggle For Russia* (New York: Random House, 1994) 276-77.

30. "The Final Tally," *The Economist* 8 Jan. 1994: 55.

31. Steven Erlanger, "Accepting Reformer's Resignation, Yeltsin Tries to Prevent Another," *New York Times* 18 Jan. 1994: A8. Steven Erlanger, "Reformer Quits New Yeltsin Cabinet," *New York Times* 21 Jan. 1994: A1. Steven Erlanger, "Russia's Reform Train Isn't Exactly an Express," *New York Times* 23 Jan. 1994: A5. Steven Erlanger, "Finance Minister Shuns Yeltsin Plea and Quits Cabinet," *New York Times* 27 Jan. 1994: A1. Thomas L. Friedman, "Russia Policy: A U.S. Riddle," *New York Times* 27 Jan. 1994: A1.

32. Claudia Rosett, "Ruble's Slide Spotlights Power of Old-Style Central Bank Chief," *Wall Street Journal* 20 Jan. 1994: 1. Steven Erlanger, "Russian Premier's Star is Rising Fast," *New York Times* 26 Jan. 1994: A6.

33. Alessandra Stanley, "Russia Congress Votes to Release Yeltsin Enemies," *New York Times* 23 Feb. 1994: A1.

34. "If Elections Were Held," *Monitor* vol. 1, no. 40, 27 Jun. 1995. "Lebed Gains, Yeltsin Loses in Presidential Preference Poll," *Monitor* vol. 1, no. 53, 17 Jul. 1995. Victor Yasmann, "The Russian Election Campaign Begins," *Prism* pt. 1, 21 Jul. 1995.

35. These figures are in 1989 and 1990 rubles, respectively.

36. Stockholm International Peace Research Institute, *SIPRI Yearbook* 1992 (Oxford: Oxford UP, 1992) 137. *SIPRI* 1994, 421-31.

37. Doug Clarke, "Hard Times in Defense Industry," *RFE/RL Daily Report* no. 221, 22 Nov. 1994.

38. "Space Center Reports on 1994 Activities," Moscow *TASS* (26 Jan. 1995) *FBIS* 30 Jan. 1995: 24. "Ekonomicheskaya Situatziya v Rossii," *Ekonomika i Zhizn'* no. 1, Jan. 1995: 1. David P. Radzanowski and Stephen J. Garber, "The National Aeronautics and Space Administration," *Congressional Research Service Report for Congress* 95-336 SPR, 1 Mar. 1995: 13. *Budget of the United States Fiscal Year* 1996 (Washington: GPO, 1995) 16.

39. Helen Gavaghan, "Soviets Search for Partners in Space," *New Scientist* 2 Nov. 1991: 17.

40. Tim Beardsley, "Selling to Survive," *Scientific American* Feb. 1993: 94. Appendix A, Chart 5 shows estimates of the Soviet GNP from 1955-1991.

41. Peter Aldhous, "Can Russia Slim Down to Survive?" *Science* 19 Nov. 1993: 1201. Penny Morvant and Thomas Sigel, "Concern About Brain Drain," *OMRI Daily Digest* no. 90, pt. 1, 10 May 1995.

42. Tim Beardsley, "Brain Drain," *Scientific American* Apr. 1992: 17. Beardsley, "Selling" 94.

43. Aldhous, "Slim" 1201. Morvant and Sigel, "Concern."

44. Western scientific associations helped restore subscriptions; pages 132-33.

45. Leif J. Robinson, "Tough Times for Astronomers in the Former Soviet Union," *Sky and Telescope* Sep. 1992: 257.

46. Beardsley, "Brain" 17.

47. Igor S. Drovenikov, personal interview, 14 Aug. 1993. Mr. Drovenikov was, at the time of the interview, Chairman of the Department of General Problems of Science and Technology Development at the Russian Academy of Sciences' Institute of the History of Science and Technology.

48. Personal interview, 4 Aug. 1993. Viktor N. Sokol'sky is author of, among other works, *A Short Outline of the Development of Rocket Research in the U.S.S.R.* (Jerusalem: Israeli Scientific Translations, 1967).

49. *SIPRI* 1992, 136. Beardsley, "Selling" 93.

50. Malcolm Gray, "The Last Soviet," *MacLean's* 24 Feb. 1992: 26-27.

51. Evgenii L. Feinberg, "Soviet Science in Danger," *Physics Today* May 1992: 34. Robinson, "Tough" 257. Beardsley, "Brain" 17.

52. Craig Mellow, "Red Rocket's Glare," *Forbes* 26 Oct. 1992: 64.

53. Morvant and Sigel, "Concern."

54. Stephen Budiansky, "A Scientific Bazaar," *U.S. News & World Report* 4 May 1992: 60.

55. Loren Graham, "Is Russian Science Dead?" Harvard Russian Research Center, Director's Seminar, 17 Nov. 1993.

56. Loren Graham, *Science in Russia and the Soviet Union* (New York: Cambridge UP, 1993) 195.

57. Aldhous, "Slim" 1201. For example, others estimated that 34,000 Russian scientists emigrated between 1989 and 1995. Morvant and Sigel, "Concern."

58. Jeremy Webb, "Bleak Future for Former Soviet Scientists," *New Scientist* 18 Apr. 1992: 7.

59. Aldhous, "Slim" 1201.

60. Malcolm W. Browne, "End of Cold War Clouds Research As Openings in Science Dwindle," *New York Times* 20 Feb. 1994: A1.

61. "Recipe for ex-Soviet Republics' Science," *Nature* 2 Jan. 1992: 2.

62. Yuri Kanin, "Exodus Likely to Increase," *Nature* 5 Dec. 1991: 343.

63. Christopher Anderson, "Russian Science Aid Falls Short," *Science* 10 Sep. 1991: 1380.

64. John Maddox, "Russian Summer Hopes Dwarf Doubts," *Nature* 4 Jun. 1992: 357.

65. Feinberg, "Danger" 36.

66. Beardsley, "Selling" 95.

67. According to Russian Academy of Sciences President Osipov. Beardsley, "Selling" 95.

68. "Russian Space Program Seeks Needed Funding," *Aviation Week* 23 Aug. 1993: 24.

69. Maxim V. Tarasenko, "Transformation of the Soviet Space Program After the Cold War," *Science and Global Security* 1994, vol. 4: 359.

70. Russian Academy of Sciences, Analytics Center. Loren Graham, "Is Russian Science Dead?" Stanford University, Donald M. Kendall Lectures, 7 Apr. 1995.

71. Graham, *Science* 190-92.

72. William Sweet, "European and U.S. Proposals for Aiding FSU Science Vie for Support," *Physics Today* Jun. 1992: 69.

73. Graham, *Science* 191.

74. Graham, *Science* 194-95.

75. Graham, *Science* 195.

76. Peter Aldhous, "New Foundation Goes Back to Basics," *Science* 19 Nov. 1993: 1200. "Battle Expands Over Shrinking Budget," *Science* 14 Jan. 1994: 166.

77. Loren Graham, telephone interview, 7 Mar. 1994.

78. As stated on page 125, while Russia's science budget shrank by two-thirds from 1990 to late 1993, its scientific work force was reduced by only thirty percent.

79. Comment is by Yuri Altukhov, Director of Moscow's Vavilov Institute of General Genetics. Aldhous, "Slim" 1202.

80. Sweet 67-68. Jeffrey Mervis, "A Way to Wean Weapons Makers," *Nature* 27 Feb. 1992: 756. Thomas L. Friedman, "U.S. to Offer Plan to Keep Scientists at Work in Russia," *New York Times* 8 Feb. 1992: A1.

81. *SIPRI* 1994, 290.

82. Yelena Abramova, "International Center to Finance Science in Russia," Moscow *TASS* 30 Jun. 1995.

83. Dan Charles, "Bargain Hunters Snap Up Russian Brainpower," *New Scientist* 14 Mar. 1992: 13. William J. Broad, "Genius for Hire: The Soviet's Best, at Bargain Rates," *New York Times* 15 Mar. 1992: E3.

84. Budiansky, "Bazaar" 59.

85. Sweet 56. Loren Graham, lecture, "Is Russian Science Dead?" Center for Russian and East European Studies, Stanford University, Stanford, California, 7 Apr. 1995. Dr. Graham was a member of the International Science Foundation's Advisory Board.

86. "American Physical Society and American Astronautical Society Launch Aid Programs for Former Soviet Union Scientists," *Physics Today* May 1992: 56. Robinson, "Tough" 258.

87. Jeffrey Mervis, "The West Gropes for Ways to Help," *Nature* 30 Apr. 1992: 733.

88. Traci Watson, "$100 Million Pledged to Support ex-Soviet Science," *Nature* 17 Dec. 1992: 617.

89. Watson, "$100 Million" 617.

90. Budiansky, "Bazaar" 59.

91. "Helping Soviet Science (continued)," *Nature* 19 Mar. 1992: 180. Loren Graham, personal interview, 5 Apr. 1994. Dr. Graham was Chairman of the relevant MacArthur Foundation committee.

92. Mr. Soros originally dedicated $100 million to aiding former Soviet science; he later matched a Russian contribution of $12.5 million. The ISF promoted the peer review system and otherwise supported would-be reformers of Russian science such as Saltykov and Vorontsov; the end of his support spelled trouble for their efforts. Loren Graham, telephone interview, 7 Mar. 1994. Graham, "Dead?"

93. Eventually the dispute over the Proton was settled by a government decree which merged Khrunichev, the manufacturer of the Proton, with its main designer, Salyut. Marcia Smith, "Russia/U.S. Space Interaction: A Trip Report With Observations and Opinions," *Congressional Research Service Report for Congress* 92-774 SPR, 27 Oct. 1992: 2-4.

94. "The Russian Space Agency is to be entrusted with....exercising the functions of a general contractor for space systems, complexes and facilities for scientific, national economic and defense purposes...." As established in the "Decree Establishing the Russian Space Agency." Printed in Nicholas L. Johnson, *Europe and Asia in Space* vol. 1991-1992 (Colorado Springs: Kaman Sciences Corp., 1994) 277.

95. Tarasenko, "Transformation," 347-48.

96. Smith, "Space Interaction" 3-4.

7

Obstacles to Cooperation in Space, Post-Cold War: International Barriers, New and Old

The end of the Cold War left Russia in turmoil. Post-Cold War cooperation in space not only had to overcome Russia's internal difficulties: it had international challenges, both new and old, to face. The new challenges arose from the fact that the dissolution of the Soviet Union left important, interdependent space and astronomy facilities in several newly independent nations. These new countries' relationships with Russia proved to be troublesome. What is more, as cooperation became more involved, the intrinsic relationship between space and militarily sensitive technologies threatened to dampen collaboration. The United States and Russia did not always agree what amounted to responsible sharing of space equipment and know-how, and what constituted the proliferation of ballistic missile and other weapons-related technologies.

The Splintering of the Soviet Space Program

The dissolution of the Soviet Union introduced yet another significant dilemma for the former Soviet space program. While the Soviet Union had exerted its centralized control over the scientific facilities spread throughout its eleven time zones, inter-republic cooperation was a non-issue. After the USSR's collapse, this potential problem became an important issue.

When the leaders of Russia, Ukraine and Belarus agreed on December 8, 1991 to dissolve the Soviet Union, they also agreed that the new Commonwealth of Independent States would maintain "joint control over nuclear weapons," but left the specifics of this control unclear.[1] On

December 21, 1991 Russia, Ukraine, Kazakhstan and Belarus signed an "Agreement on Joint Measures with Respect to Nuclear Weapons." In this accord all four nuclear states agreed to pursue ratification of the first Strategic Arms Reduction Treaty (START 1), to form a joint nuclear policy and to give the Russian president the power to decide to use nuclear weapons with the consent of the other three states' leaders. In addition Ukraine and Belarus agreed to transfer all of their nuclear weapons to Russia for destruction and to accede to the 1968 nuclear non-proliferation treaty as non-nuclear states. That same day all eleven members of the newly formed Commonwealth of Independent States (CIS) agreed to maintain a unified command of strategic military forces.

What the heads of state did not define was what they meant by "strategic military forces." Nine days later the eleven former republics that joined the CIS met again, signing an agreement defining "strategic" as including the former Soviet strategic missile forces, air force, navy, air defenses, directorate of the space command, airborne troops and strategic and operational intelligence.[2]

On February 14, 1992 Ukraine, Azerbaijan and Moldova, fearing what they saw as increasingly imminent Russian domination, refused to sign further accords. A month later Ukraine refused to sign an accord on space. After it became clear that the Commonwealth of Independent States would form no strong joint military structure, the Russian Defense Ministry reclaimed control over the Russian military space program in August 1992. Thus began the inter-former republic disputes.

These disputes posed new problems for the Russian space program. Although 80-90% of former Soviet space contracts were in Russia,[3] 75-90% of the former Soviet space industry was located in Russia[4] and an estimated 80% of former Soviet space "assets" were Russian,[5] Russia faced significant handicaps in its ability to carry on the Soviet Union's space program. Some of the Soviet Union's ground bases, tracking stations and early-warning radar systems left with the Baltic republics. In addition, inter-former republic border disputes between Russia and Ukraine, Azerbaijan and Armenia and Russia and Georgia threatened the Crimean Astrophysical, Byurakan and Ratan-600 observatories, respectively.[6]

Russia and Ukraine

The most significant challenges, though, involved relations between Russia and Ukraine and Russia and Kazakhstan. The newly sovereign Ukraine had a population of 51.9 million people and 1,804 nuclear warheads.[7] Ukraine was host to the important Yuzhnoe industrial complex. The Yuzhnoe complex held an essential monopoly over Soviet production of the Tsiklon and Zenit rockets, which were an important part of the Soviet Union's space program. The Tsiklon was used to launch scientific, photographic reconnaissance and electronic intelligence satellites; the Zenit was a relatively new, highly automated booster used to fill the Soviet need for launching large, low-altitude electronic intelligence satellites.[8]

Yuzhnoe and other Ukrainian space concerns also held an essential monopoly over technologies critical to electronic intelligence, early-warning and radar ocean reconnaissance satellites. In 1992 experts estimated that Russia would not be able to replicate Ukrainian satellite technologies until the late 1990s. The loss of space-based early warning capacity would have been especially troublesome for Russia, since it was already faced with the dilemma of finding many of the Soviet ground-based warning systems now belonging to other countries. In addition, other Ukrainian enterprises provided subsystems important to other types of satellites and aspects of the space program. For example, Ukraine's NPO Hartron was a major supplier of computers, control systems and software for rockets; its systems were integral parts of almost all Soviet ICBMs.[9] Finally, Ukraine possessed one of the Soviet Union's most important astronomical facilities, the Main Astronomical Observatory of Ukraine. Thus, in order to come close to maintaining the Soviet space program Russia and Ukraine needed to cooperate, at least to some degree.

Just as American and Soviet/Russian space relations were/are a subset of the larger bilateral relationship, so, too were Russo-Ukrainian efforts to cooperate in space subject to overall relations between the two nations. Russo-Ukrainian tension in the early 1990s was reflected and exacerbated mainly by five interrelated issues: the transfer of nuclear weapons, the Crimea, the former Soviet Black Sea Fleet, oil payments and Ukraine's desire to associate with the North Atlantic Treaty Organization (NATO).

First, Ukraine grew increasingly weary of Russian expansionism and acted extremely slowly to fulfill its agreement to become a non-nuclear state. According to the U.S.-Soviet Treaty on the Reduction and Limitation of Strategic Offensive Arms (START 1) and the 1992 Lisbon Protocol Ukraine was to transfer all of its 176 strategic nuclear missiles to Russia for destruction. However, In 1993 Ukraine declared itself the owner of all strategic nuclear warheads deployed or stored on its soil, and proclaimed that the 46 SS-24 missiles on its territory did not fall under previous agreements. Russia was working to consolidate itself as the sole nuclear successor to the Soviet Union and protested this declaration vigorously.[10] On November 18, 1993 the Ukrainian parliament finally ratified the START 1 treaty but declined to agree to become a non-nuclear state. Instead the parliament agreed to "gradually" destroy its nuclear arms after a list of thirteen conditions, centering around compensation and security guarantees, were met.[11]   On January 14, 1994 Presidents Clinton, Yeltsin and Kravchuk signed yet another accord resolving that Ukraine would transfer its nuclear warheads to Russia and accede to the nuclear non-proliferation treaty as a non-nuclear state. Even then the Ukrainian parliaments' approval remained uncertain. Finally, on March 5, 1994, Ukraine made its first shipment of nuclear weapons to Russia, transferring sixty warheads for destruction.   In November 1994 the Ukrainian parliament, having finally received the security and financial assurances it wanted from the U.S. and Russia, ratified the nuclear Non-Proliferation Treaty as a non-nuclear state. This placed Ukraine firmly on the path of de-nuclearization.[12] As a reward for this de-nuclearization, the presidents of the United States and Ukraine expanded U.S.-Ukrainian cooperation in space.

Second, in 1954 Khrushchev, in a controversial decision, transferred the Crimea, a peninsula on the Black Sea, to Ukraine.   Ethnic Russians comprised approximately seventy percent of Crimea's population.   The dissolution of the Soviet Union renewed this long contentious issue, as some Russians demanded that Crimea be re-joined to Russia.[13]   These demands were fueled by the fact that Ukraine's economy was in much worse condition than Russia's.   In 1992 Ukraine granted Crimea special status as an autonomous republic, but this only provided pro-Russian Crimeans with a medium for gaining further concessions from Ukraine's central government.[14]   In a January 17, 1994 election a pro-Russian Crimean candidate won the largest plurality, forty percent of the vote, while the only candidate advocating that Crimea remain in Ukraine won only sixteen percent.[15] Crimea's future remained uncertain through early 1994, but the new Ukrainian government led by Leonid Kuchma (the former prime minister and former director of NPO Yuzhnoe) managed to curry more Crimean support beginning in mid-1994.   Crimea, with Ukrainian support, began to demand compensation for environmental damage it had suffered as a result of the stationing of the Black Sea fleet on its territory.   Whatever the future of Crimea, it remained a thorn in the side of Russo-Ukrainian relations in the mid-1990s.

Crimea was largely a problem because it was host to the third and perhaps most contentious issue between Russia and Ukraine: it was the home port of the Black Sea Fleet. Although Crimea was Ukrainian soil, Russia claimed the fleet as its own.[16]   From almost the day the Soviet Union dissolved Russia and Ukraine went through round after round of negotiations over the fleet. On several occasions the two sides announced agreements over how to divide up the fleet and how to compensate Ukraine for its smaller share, etc., but each of these agreements broke down when the two sides tried to iron out the details.   For example, in June 1995 Presidents Yeltsin and Kuchma announced that they had reached an agreement, whereby the fleet would be evenly divided but Russia would purchase about sixty percent of Ukraine's share. However, by the time the nations' prime ministers were due to meet in August in order to iron out the agreement's details, the two sides had so antagonized each other that the meeting was canceled. Still, some progress was apparent.

The fourth problem between Russia and Ukraine was partly related to the Black Sea fleet. In early February 1994 Ukraine announced that it would join NATO's cooperative Partnership for Peace program, against Russian wishes. The North Atlantic Treaty Organization proposed the Partnership for Peace as a form of associate membership for Central and Eastern European nations at a January 10-11, 1994 conference in Brussels, and Ukraine was the fourth former Soviet republic (after the Baltic republics), to seek membership in the cooperative plan.[17] Although Russia was offered a "special relationship" with NATO, Russian politicians pushed for more Euro-centric security organizations, arguing that NATO was outdated since the Cold War was over. However, this rhetoric did not stop Ukraine from trying to balance its enormous neighbor to the north by establishing closer ties with the Atlantic alliance. In a particularly poignant gesture, Ukraine participated in joint naval exercises with the United States, Bulgaria, Romania, Turkey, Greece, Italy, France and Spain in the Black Sea even as the August 1995 ministerial meetings to negotiate the future of the Black Sea fleet broke down.

Finally, Ukraine consumed about 250 million cubic meters of Russian gas each day in early 1994. However, in early March 1994 Russia, angry over Ukraine's failure to pay the more than $1 billion it owed Russia, began cutting off Ukraine's fuel supplies.[18] This particular sticking point was largely solved after March 1995, when Ukraine managed to reschedule its debt with Western assistance. After this Ukraine was able to make the majority of its fuel payments to Russia on time.[19]

Logical Partners in Space

In sum, Russian-Ukrainian relations were problematic in the first half of the 1990s. Thus, it is not surprising that each of the two countries attempted to make their space programs more independent from the other's. Russian space officials were often frustrated with their dependence on Ukrainian subsystems. As Lieutenant General Vladimir Durnev, deputy commander of Russia's Strategic Space Forces, told the Russian news agency in early 1995, "Russia must pursue a space policy of its own, fully independent of any other state."[20] Russian-Ukrainian disagreements helped delay the launch of a Photon-10 microgravity research satellite. The Photon-10 mission was a cooperative effort with the European Space Agency and the French space agency, and was originally scheduled for launch in September 1994 on a Soyuz booster. However, Russia and Ukraine were unable to agree upon prices for Ukrainian components critical to the Soyuz launch vehicle and for Russian utilization of the Ukrainian Evpatoria tracking site (which is located in Crimea). As a result the launch did not take place until February 1995, and when it did, Russian space officials proudly and loudly declared that this particular booster had less Ukrainian subsystems than any previous Soyuz. In explaining the Russian announcement, Valeriy Abrashkin, deputy chief of the Russian design office responsible for Soyuz rockets, said, "This is not only our company's policy, it is the policy of the government."[21]

Russia also tried to control Ukraine's entry into the commercial launch market. After NPO Yuzhnoe managed to gain a contract with Space Systems/Loral to launch thirty-six communications satellites in three launches, Koptev, the director of the Russian Space Agency, protested: "We have an agreement with Ukraine that we are supposed to negotiate the terms of all commercial launches....That was not done for this contract, and still has not been done." Russia's authority over the launch was basically derived from the fact that the Zenit had to be launched from Baikonur, which the Russians had leased from Kazakhstan (see below). The contract became the subject of a heated public debate, and proponents of the more-Russian Soyuz booster publicly exhorted Loral that the Zenit was a poor choice, as based upon "perfect knowledge of the Russian market, space, and the geopolitical overlap of the different CIS states [a euphemism for Ukraine]."[22] Russia and Ukraine also competed in proposals to establish launch sites in Australia.[23]

Despite these strains in Russo-Ukrainian relations, Ukraine and Russia continued to cooperate more than they competed in space in the mid-1990s, because of the simple fact that they needed each other. Ukraine had a vested interest in maintaining its space infrastructure, and this required capital. It had products and services to offer, but no launch site. Ukraine was also dependent upon Russia for oil and gas and realized that Russia was militarily dominant.[24] Russia, in turn, needed Ukrainian Russian components, craft and vehicles. Thus, Ukraine continued to fulfill the vast majority of its space obligations to Russia, and, in the summer of 1992, it finally signed the space addition to the December 1991 Minsk accords. In the fall of 1994 Russia and Ukraine reached an agreement that paved the way for the first launch of a Russian military satellite on a Zenit booster, and on November 24, 1994, a Zenit launched a Russian Electronics Intelligence satellite. In addition, Russia and Ukraine began exploring the use of Zenits to resupply space station Mir.

It is likely that, given both states' mutual interests, Ukraine and Russia will continue to cooperate more than compete in space.

Russia and Kazakhstan

Kazakhstan presented the other great challenge to Russia. The Baikonur Cosmodrome, one of the Soviets' two main launch pads, is located near Leninsk, Kazakhstan. Baikonur launched every piloted Soviet flight and is where the Energia rocket and Soviet space shuttle Buran were built. Russia possesses no corresponding launch site; the Russian Plesetsk Cosmodrome is restricted by the fact that it is much farther north (thus reducing payload capability for most orbits) and cannot launch Proton rockets.[25] Russia acknowledged Kazakhstan's important role in the NIS's space program in July 1991 when it gave in to Kazakh pressure, changing its piloted launch schedule to accommodate a Kazakh cosmonaut for launch to Mir in October of that same year.[26]

Although Baikonur is located in Kazakhstan, it was in fact operated and staffed dominantly by Russians. As of late 1993 the cosmodrome was staffed by 8,000 Russians, while only thirty-eight Kazakhs served in rocket units (other Kazakhs held maintenance, and other non-technical jobs, including a 120-man militia).[27] According to the December 30, 1991 Minsk agreement, Baikonur was to be controlled by the Interstate Space Council, which Russia dominated. However, as the design for a CIS military command dwindled, so, too did the idea of an Interstate Space Council die.[28] Even before the December Minsk agreement, in October 1991, General Ivanov, the commander of the Russian Military Space Force (RMSF) aggravated the issue by unilaterally declaring Baikonur to be under his control. Kazakhstan protested and founded its own Space Research Agency; on December 20, 1991 the RMSF responded with a show of force, launching an SS-19 ICBM--without notifying or seeking the approval of Kazakhstan's government.[29] This launch stimulated a series of bitter public arguments over who was responsible for upkeep of the facility.[30] For example, in January 1993 the RMSF claimed it had spent one billion rubles (a figure that may have been an exaggeration) on Baikonur upkeep and assailed Kazakhstan for not playing its role. In mid-April 1993 Yeltsin demanded a personal meeting with Kazakhstan's President Nazarbaev to discuss the growing conflict.[31]

In March 1994 Ivanov announced that Russia was planning to convert a decommissioned missile base in the Russian Far East into a cosmodrome. The site was near Svobodny, close to the Russian-Chinese border and at roughly the same latitude as Baikonur. However, it was never clear if Russia seriously intended to undertake this extremely expensive operation, or if announcements such as Ivanonv's were intended to soften Kazakhstan's negotiation stance. President Yeltsin personally estimated that it would cost $2 billion to convert the former missile site into a space launch base.[32] In another interesting twist, one which brought Ukraine into the debate, some Russians suggested that Russia could finish upgrading Plesetsk to be able to launch Zenit rockets. The upgrades were almost complete until they were suspended shortly after the dissolution of the Soviet Union, and could allow cosmonauts to be launched from the Russian cosmodrome.[33] In another attempt to lessen its dependence on Baikonur, the Khrunichev State Research and Production Center began designing a heavy-lift booster, Angara, which was intended to fill many of the needs its Proton booster fulfilled from Baikonur.

Despite all of these difficulties, the mutual need between Russia and Kazakhstan's space programs is even stronger than that between Russia and Ukraine's. Russia's space program would be greatly hampered without Baikonur, and Kazakhstan has essentially no space program other than Baikonur. Russia needs the launch facility, and Kazakhstan has none of its own rockets or satellites to boost from the cosmodrome. As Major General Vyacheslav Bezborodov, deputy chief of staff of the Russian Military Space Forces put it,

> The question is this: Does Russia need Baikonur? In other words, are the tasks that we are now performing using the launchers stationed in Kazakhstan essential? These tasks include early warning, communications, television, monitoring compliance with the terms of conventional and nuclear arms limitation treaties, navigation, long-distance probing of the Earth, and manned space exploration. There is no alternative--we cannot do without Baikonur....[34]

Thus, even before early 1994 Kazakhstan allowed Russia to continue to use Baikonur on a commercial basis; again, the laws of supply and demand predicted this collaboration. Under the agreement, Russia was responsible for 94% of the expense of maintaining the cosmodrome, while Kazakhstan paid the remaining 6%. In the meantime both sides worked to solidify this somewhat ad hoc solution. Finally, on March 28, 1994 Russia and Kazakhstan signed a long-term lease granting Russia control of the cosmodrome and its associated facilities for twenty years. In exchange Russia agreed to pay Kazakhstan $115 million annually and to contribute to Leninsk's socio-economic infrastructure. The agreement, interestingly enough, transfers the cosmodrome to the Russian Space Agency, not to the Russian Military Space Forces, although the cosmodrome was in fact to be staffed by RMSF troops. These roughly 16,000 soldiers were to be under contract to the space agency.[35] Russia seemed eager to stress the Baikonur's peaceful applications and that Baikonur was not becoming a Russian military base. As Ostankino television reported,

> Boris Ostroumov, Deputy General Director of the Space Agency, assured journalists that Baikonur will not become a Russian military base. Most of the facilities will be used for scientific and national economic needs, and will be administered by the Russian Space Agency.[36]

Whatever these semantics, the accord was implemented, after some additional haggling, by both governments: in July 1994 the Kazakh parliament ratified the agreement and the Russian Duma approved the corresponding law; in September 1994 Prime Minister Chernomyrdin decreed the implementation of the accord; Yeltsin built upon this decree with an October 1994 edict; and in April 1995 the Russian Duma ratified what had by then become a treaty. Unless either side objects, the lease can be extended for another ten years.[37]

A July 1994 Baikonur launch to Mir was an encouraging sign for Russian-Ukrainian and Russian-Kazakh cooperation in space. The July 1994 Soyuz launch was the first mainly Russian piloted flight which did not carry a citizen of Russia. One of the pilots was a Kazakh citizen, the other a Ukrainian. Kazakh President Nazarbaev and Russian Deputy Prime Minister Oleg Soskovets attended the launch, and afterwards held a joint press conference where they praised the virtues of collaboration in space.[38]

A Withering Cosmodrome?

However, in this case the mutual benefits may not be enough to facilitate U.S.-Russian/Kazakh collaboration on the level that could have been possible between the United State and the Soviet Union. Baikonur, which sprawls across 600 square kilometers in the dry steppes near the Aral Sea, experiences temperature ranges from -40 to 122 Fahrenheit; this extreme climate causes structural damage to the spaceport's infrastructure and mandates regular maintenance. Because of Russia's and Kazakhstan's financial woes and because the two states disagreed over who controlled Baikonur, the cosmodrome was allowed to deteriorate. Various reports described Baikonur's buildings as sinking into the ground, their roofs pocked by gaping holes, its launch pads stripped by vandals and its communication lines dilapidated to the point of unreliability.[39] Occasionally but not infrequently water and power supplies were cut off. According to one account, one hour before the Russians launched the piloted *Soyuz TM-17* on July 1, 1993, the power went out at the launch site. Only emergency measures resupplied power to the cosmodrome, and, immediately after the launch, the entire city of Leninsk was without power. Colonel Yevgeni Aleksandrov, one of Baikonur's two senior commanders, said after the July 1 launch that, "Space is filled with surprises. But the biggest surprise of all was that the takeoff of *Soyuz TM-17* was successful."[40]

Worrisome reports continued after the July 1993 electrical failure. In July 1994 a working group of the Russian Duma's Committee for Industry, Construction, Transportation and Power Engineering returned from an assessment trip to Baikonur. They described the situation at the cosmodrome as "disastrous." On March 7, 1995 there was a fire at a Baikonur ground processing facility. The Russian news agency reported that staff had a hard time extinguishing the fire because Russia had recently reduced shipments of fire-fighting equipment to Baikonur because Kazakhstan was unilaterally declaring the equipment its property as soon as it entered the country.[41]

Meanwhile Baikonur's workers grew increasingly dissatisfied with their low pay and poor standard of living. On February 23, 1992 Baikonur's workers rioted, complaining of low pay and ill treatment; at least three died.[42] On June 3, 1993 Russia's Leninsk soldiers rioted, burning down three barracks, three headquarters, a club, a hospital and a library.[43] The Duma Committee which visited the space center in the summer of 1994 observed that

> the situation in the city of Leninsk...is on the verge of a social explosion....The houses are falling apart, the crime situation is growing worse, schools kindergartens and hospitals are closing down.[44]

As though to reinforce their warnings, shortly after the Committee left the area a gas explosion occurred in Leninsk apartment buildings, killing at least fourteen people and damaging at least nineteen apartment buildings. The Russian news agency commented:

> The ultimate cause of the tragedy consists in the poor conditions of daily life in Leninsk, experts say. Engineering communications, power transmission lines need to be replaced urgently. Most of the apartment buildings in Leninsk are in a state of disastrous disrepair.[45]

Thus, not surprisingly, Baikonur's ability to continue to provide safe launches became a controversial issue. NASA and Department of Defense officials made several trips to survey the cosmodrome, and in the fall of 1993 the Clinton administration asked the private consulting firm ANSER to visit the site and review its deterioration. The September 1993 ANSER report suggested that although they found a "marked deterioration in living standard, a reduction in operations and personnel and a military unwilling to give up control to a civilian agency," Baikonur continued "to successfully launch a wide variety of rockets from reliable facilities." However, the ANSER team's ability to evaluate Baikonur's state was questioned by Aleksandr Shumilin, the cosmodrome's Chief, in an open magazine article:

> What is the true state of affairs? In an attempt to answer this question, representatives of the U.S. firm of ANSER...visited Baikonur....After spending less than two days at the cosmodrome and viewing the launch and technical complexities of the Soyuz, Proton and Energia rockets (30 minutes at each complex), the guests happily set off home. Not even a high-class specialist is able to assess the technical condition of very complex systems in such a short space of time, apart from forming an impression of their outward appearance. But the U.S. experts drew hasty conclusions....Meanwhile, the U.S. experts maintain that the tragic situation...has been 'greatly and deliberately' exaggerated.

Elsewhere in the same article Shumilin argued, "Baikonur really is in need of serious assistance to maintain impeccable technical readiness."[46]

Baikonur's fate is essential to the future of U.S.-NIS cooperation in space, and therefore also to the United States' entire space station effort. Without Baikonur Russia would lose the home to at least forty percent of its launches, including all of its piloted missions. As Marcia Smith stated in testimony to the House of Representatives' Committee on Science, Space and Technology on October 6, 1993, conditions at the Baikonur cosmodrome were "deteriorating, raising concern about the health of the physical infrastructure needed to accomplish the program [of cooperation]." Clinton officials, who had tied their foreign policy success to joint achievements in space, had reason to be hopeful: on December 15, 1993 Vice President Gore described Baikonur as "superb, on par with the best launch facilities anywhere in the world."[47] Indeed, if the United States and Russia were to fulfill the agreements that he and Chernomyrdin signed on September 2, 1993, Baikonur would have to remain usable well into the twenty-first century.

By late 1993 it seemed that Baikonur's situation was growing increasingly dire. Congress was more skeptical of the executive's optimistic assessment of the situation, and Representative George Brown (D - California), chairman of the House Science, Space and Technology Committee, which authorized funds for U.S. space station spending, led a delegation of seven representatives to Baikonur in December 1993. The delegation found "no substantive evidence of neglect or deterioration," but noted that "major upgrade and investment" would be needed to guarantee the facilities' viability. Especially troublesome, the group concluded, was the surrounding area's socio-economic conditions: "At a minimum," their report stated, "restoration of utilities, municipal services and food distribution to levels of two years ago would seem necessary."[48]

Although reports were varied, by this time there was a consensus of evidence that Baikonur was functioning but teetering too close to the edge of disaster for comfort. Thus, the March 1994 Russian-Kazakh agreement was a well-timed major boost. The treaty provided specific funds for upgrades to Baikonur and the surrounding area's infrastructure. In mid-1994 the Lockheed-Khrunichev-Energia joint venture (Chapter Eight, pages 163-64 and 166) announced that it had agreed to provide funds to maintain and upgrade Baikonur's facilities, including systems for processing payloads, personnel accommodations and communications equipment. The joint venture's board of directors stated that the Russian side (i.e., Khrunichev and Energia) had already spent some $1.5 million on these kinds of upgrades by August 1994, suggesting that their announcement boiled down to the fact that the American space giant Lockheed Martin was committing significant funds to the cosmodrome.[49] It was not clear if LKEI's expenditures were a result of the Russian-Kazakh agreement creating a more stable environment for investment, and/or if the joint venture was trying to protect previous investments. Whatever the case, these developments were followed by evidence that Baikonur was on an upswing. Launch rates were generally on schedule; in fact, despite all the controversy, Baikonur hosted thirty launches in 1994, all of which were successful, in comparison with America's Cape Canaveral and Vandenberg Air Force Base, which together launched twenty-four boosters successfully.[50] As ANSER officials observed after a summer 1995 return visit to the cosmodrome, "the condition of the operational facilities is very good but not up to Western standards for cleanliness and appearance."[51]

Perhaps the U.S. and NIS space programs would have to come closer to a common understanding of cleanliness, but as long as Baikonur could successfully serve as a busy spaceport, the future boded well for cooperation. Essentially, in the first half of the 1990s the cosmodrome's essential facilities were being maintained, but everything else was being allowed to deteriorate--a formula which, if followed for too long, would lead to disaster.

A Pause in Cooperation: Launch Vehicles Versus Missiles

The end of the Cold War did not change the fact that space and nuclear weapons are inextricably intertwined. After the dissolution of the Soviet Union the United States continued to promote non-proliferation efforts.

America's proliferation concerns clashed with Russia's need for income when Russia sought to sell India cryogenic rocket engines. Russia and India insisted that the technology transfers were aimed at aiding India's indigenous satellite launching program, but the United States was concerned that the technology could be applied to ballistic missile development. As Russia moved forward with the sale in May 1992, the Bush administration imposed a two-year ban on exports and imports to and from the Russian Glavkosmos (the organization responsible for the sale) and the Indian Space Research Organization. Despite the sanctions, Yeltsin approved the deal in early 1993.[52] This disagreement was settled in July 1993, when Yeltsin agreed to reduce the amount of technology transfers involved by transferring the engines but not the information required to build them. The compromise met no party's original goals, but did satisfy the United States' basic concerns and thereby fulfill an American prerequisite for the September 1993 accord merging space stations Alpha and Mir 2. When Gore and Chernomyrdin met that September in Washington, they signed a bilateral agreement adhering Russia to the Missile Technology Control Regime.[53] The sanctions were allowed to expire in May 1994, but after that Russian transfers of rocket engine technology to Brazil and China became controversial.[54] Although none of these disagreements halted space cooperation, Russia's growing need for hard currency, the continuing spread of missile technology and America's ongoing proliferation concerns suggested that this contentious issue was likely to resurface in the future.

The Reshuffling at the End of the Cold War

Although the end of the Cold War changed many of the factors affecting U.S. space policy, Russian space policy and U.S.-Russian space cooperation, and some Cold War policy influences persisted into the mid-1990s. In addition, although the end of the superpower rivalry facilitated greater levels of U.S.-Russian space cooperation than were previously possible, other changes occurring simultaneously threatened to undermine the cooperation thus facilitated. Russia's economic crisis and brain drain, its space industry's dis-organization, the Russian Academy of Sciences's resistance to change and intra-CIS tensions all made U.S.-Russian collaboration in space more difficult. An improved international climate is a precondition for extensive cooperation in space, but does not dictate that such cooperation will actually occur or be successful.

During the early 1990s the West offered Russian science significant aid in the form of direct grants and mutually beneficial contracts. NASA's $400 million contract with Russia will aid the Russian space program significantly.[55] However, in the long-run, for Russia to maintain itself as a viable partner for collaboration in space, it will have to succeed on its own.   Contracts and aid packages will supplement Russia's scientific community, but eventually Russia will have to improve its economy, fund its scientists and engineers, reorganize its space and scientific communities, maintain cooperative arrangements with Ukraine and Kazakhstan and avoid unilateral technology transfers which threaten to spread missile technologies--on its own--in order to continue to facilitate cooperation in space with the United States.   Finally, returning to the fundamental precondition which facilitated the consideration of all of these issues, United States-Russian cooperation in space will continue to be subject to stable, friendly overall U.S.-Russian relations.

Notes

1. Chapter Five, pages 99-100.   "Agreements Establishing the Commonwealth of Independent States," *International Legal Materials* vol. 31, no. 1, Jan. 1992: 144.

2. "Agreements Establishing" 152-53.   "Alma Ata Declaration," *International Legal Materials* vol. 31, no. 1, Jan. 1992: 148-49.

3. Gavaghan, "Search" 17.

4. *SIPRI* 1992, 139.

5. Michael D. Lemonick, "Space Program for Sale," *Time* 16 Mar. 1992: 55.

6. Robinson, "Tough" 257.

7. Douglas Jehl, "Ukraine: A Nuclear Power, but Untested Loyalties," *New York Times* 2 Dec. 1993: A1.   Paul Mann, "U.S. Wins Accord on Kiev Warheads," *Aviation Week* 17 Jan. 1994: 22-23.

8. The Zenit, which was first launched in 1985, was also originally intended as a strap-on booster for the Energia heavy launch vehicle and the Buran space shuttle. *SIPRI* 1992, 140.

9. "U.S.-Ukraine Joint Effort," *Aviation Week* 5 Dec. 1994: 54.

10. *SIPRI* 1994, 180-81.

11. "Ukraine Ratifies the Missile Pact, but Delays Ending Nuclear Status," *New York Times* 19 Nov. 1993: A10.

12. Serge Schmemann, "On Russian TV, Clinton Backs Reforms," *New York Times* 15 Jan. 1994: A1.   "Sixty Warheads Leave Ukraine to be Dismantled in Russia," *New York Times* 6 Mar. 1994: A10.

13. Robert Legvold, "Foreign Policy," *After the Soviet Union*, ed. Timothy J. Colton and Robert Legvold (New York: W. W. Norton, 1992) 162.

14. Celestine Bohlen, "For Russia and Ukraine, Crimea is a Sore Nerve," *New York Times* 23 Mar. 1994: A3.

15. "Crimea Will Hold a Runoff: Rivals Vie on Region's Fate," *New York Times* 18 Jan. 1994: A10.

16. Colton and Legvold, *After* 128-29.

17. "Ukraine Joining Plan for NATO Partnership," *New York Times* 6 Feb. 1994: A8.   Roger Cohen, "NATO and Russia Clash on Future Alliances," *New York Times*

10 Dec. 1993: A15. Steven Erlanger, "Rightist Vote Helps Russia Fight Expansion of NATO," *New York Times* 29 Dec. 1993: A1.

18. "Russia Cutting Fuel to Neighbors," *New York Times* 4 Mar. 1994: A1. Steven Greenhouse, "Clinton Vows to Improve Relations With Ukraine," *New York Times* 5 Mar. 1994: A5.

19. Vladimir Socor, "Prime Ministers Highlight Russian-Ukrainian Differences," *Prism* pt. 2, 4 Aug. 1995.

20. "Lack of Funds Threatens Future Space Launches," Moscow *TASS* (14 Mar. 1995) *FBIS* 15 Mar. 1995: 29-30.

21. Peter B. de Selding, "Russia Distances Space Program From Ukraine," *Space News* 20-26 Feb. 1995: 3.

22. Peter B. de Selding, "RKA Chief Blasts Loral's Zenit Deal," *Space News* 19-25 Jun. 1995: 1. The statement is from, Paul Aouizerate, "An Open Letter to the President of Globalstar," *Space News* 5-11 Jun. 1995: 5 [a paid advertisement].

23. "Russia, Ukraine Compete Over Australian Space Complexes," Moscow *TASS* (23 Feb. 1995) *FBIS* 23 Feb. 1995: 1.

24. Jehl, "Untested" A1.

25. Engineering Plesetsk to launch Proton rockets would cost, according to one estimate, roughly a billion 1991 rubles. "Future of Space Complex, Program Reviewed," Moscow *Izvestia* (4 Oct. 1991) *FBIS* vol. 3: 198; 11 Oct. 1991: 2-3. Estimate by Yuri Semenov.

26. V. Kiernan, "Soviets Cancel Fall Flight to Mir Station," *Space News* 29 Jul. 1991: 3.

27. Sergei Leskov, "Notes From a Dying Spaceport," *Bulletin of the Atomic Scientists* Oct. 1993: 40, 43. Mr. Leskov was chief correspondent for the Russian newspaper *Izvestia* when he wrote this article.

28. Marcia Smith, telephone interview, 26 Apr. 1993. Marcia Smith is a Congressional Research Service specialist in science and technology policy.

29. *SIPRI* 1992, 136.

30. Craig Covault, "Russians Locked in Struggle for Space Program Control," *Aviation Week* 1 Feb. 1992: 57-59.

31. Marcia Smith, telephone interview, 26 Apr. 1993.

32. "Svobodny Upgrades Pegged at $2 Billion," *Space News* 11-17 Jul. 1994: 2.

33. "Baikonur 'Will Not Be Needed' In a Few Years," *Moscow Ostankino Television First Channel Network* (28 Aug. 1994) *FBIS* 31 Aug. 1994: 36-37.

34. Valeriy Baberdin, "Does Russia Need Baikonur?" *Krasnaya Zvezda* (22 Oct. 1994: 4) *FBIS* 25 Oct. 1994: 26-27.

35. "Official Says 16,000 Servicemen to Staff Baikonur," Moscow *RIA* (27 Dec. 1994) *FBIS* 29 Dec. 1994: 4.

36. "Baikonur 'Will Not Become a Russian Military Base,'" (29 Dec. 1994) *FBIS* 10 Jan. 1995: 5.

37. In July 1994 Kazakhstan tried to up the ante to $480 million in annual rent, but it eventually conceded for the original $115 million plus satellite communications and environmental monitoring services. "Parliament Adopts Budget, Ratified Baikonur Agreement," *Almaty Kazakh Radio Network* (14 Jul. 1994) *FBIS* 15 Jul. 1994: 52. "Law Ratifies Baikonur Utilization Agreement," *Rossiyskaya Gazeta* (27 Oct. 1994: 6) *FBIS* 31 Oct. 1994: 32-33. "Government Decree on

Baikonur Space Center," *Rossiyskaya Gazeta* (7 Sep. 1994: 4) *FBIS* 8 Sep. 1994: 29-30. "Yeltsin Issues Edict on Use of Baikonur Cosmodrome," Moscow *TASS* (24 Oct. 1994) *FBIS* 25 Oct. 1994: 25-26. "Duma Ratifies Treaty on Renting Baikonur," Moscow *TASS* (21 Apr. 1995) *FBIS* 24 Apr. 1995: 32. Dennis Newkirk, "Cosmodrome News," *Countdown* Sep./Oct. 1994: 26.

38. The Kazakh was an officer in the Russian Air Force. Vladimir Li, "Baikonur, Leninsk Difficulties Evaluated," *Kazakhstanskaya Pravda* (5 Jul. 1994) *FBIS* 9 Aug. 1994: 31-32.

39. For example, Vladimir Yaropolov, who worked at Baikonur for thirteen years before transferring to the Cosmonaut Training Center in Star City, wrote, "the Baikonur cosmodrome is dying--the local hooligans are looting apartments, are dragging away kilometers of cables, pipes and conduits, and structures are falling down due to lack of proper care." "Tombstone Over Zvezdny?" *Stolitsa* Dec. 1993: 13-15. *JPRS* 3 Mar. 1994: 1.

40. Leskov, "Dying" 40.

41. Vincent Kiernan, "Russia Plans New Cosmodrome Outside Kazakhstan," *Space News* 21-27 Mar. 1994: 8.

42. "Three Killed in Space Center Riot," *Facts on File* 16 Apr. 1992: 266.

43. Leskov, "Dying" 43.

44. "Duma Deputies Term Baikonur Situation 'Disastrous,'" Moscow *TASS* (13 Jul. 1994) *FBIS* 15 Jul. 1994: 22.

45. "Accident Reveals 'Disastrous' Conditions at Baikonur Town," Moscow *TASS* (28 Jul. 1994) *FBIS* 29 Jul. 1994: 33.

46. In *Rossiyskaya Gazeta* 1 Feb. 1994. As cited in "A View from the Cosmodrome," *Washington Times* 10 Mar. 1994: A20; and "Baikonur Chief Blasts U.S. Government Study," *Space News* 14-20 Feb. 1994: 2.

47. Marcia Smith, "Testimony to the Committee on Science, Space and Technology," *Congressional Research Service Report* 6 Oct. 1993: 3. Bretton S. Alexander, Tom Cremins and James Oberg, "Exchange: The State of U.S.-Russian Space Cooperation," *Washington Times* 6 Jan. 1994: A19.

48. Ben Iannotta, "Delegation Reports Baikonur Cosmodrome in Good State," *Space News* 4-10 Aug. 1994: 12. "U.S. Team Recommends Cosmodrome Upgrades," *Aviation Week* 28 Mar. 1994: 27.

49. "Joint Venture Upgrades Baikonur Facilities," *Segodnya* (9 Aug. 1994: 3) *FBIS* 9 Aug. 1994: 41. "LKEI Board Approves Baikonur Upgrade Plan," *Business Wire* 6 Dec. 1994.

50. From 1957-1994 Baikonur hosted 971 successful launches; Plesetsk 1,409; Vandenberg 502; Cape Canaveral 492; and Kapustin Yar 83 (none in 1994). Marcia S. Smith, "Space Activities of the United States, C.I.S., and Other Launching Countries/Organizations: 1957-1994," *Congressional Research Service Report for Congress* 95-873 SPR, 31 Jul. 1995: 114.

51. James R. Asker, "U.S. Firm Says Baikonur Improving Slowly," *Aviation Week* 31 Jul. 1995: 26.

52. Sanjoy Hazarika, "Despite U.S., Yeltsin Backs Rocket Deal With India," *New York Times* 30 Jan. 1993: A2. Barbara Crossette, "Russia's Rocket Deal With India Leads U.S. to Impose Trade Bans," *New York Times* 12 May 1992: A8.

53. "U.S., Russia Settle Export Disagreement," *Aviation Week* 26 Jul. 1993: 27.

54. Bill Gertz, "Russia Sells Rocket Motors to China," *Washington Times* 13 Feb. 1995: A4. Warren Ferster, "Russia Did Not Violate MTCR, Official Says," *Space News* 12-18 Jun. 1995: 12.

55. Chapter Five, page 108.

8

The Market Bridges the Gap: Commercial Space Cooperation

Chapters Six and Seven discussed many of the problems that the disaggregation of the former Soviet space program caused. However, as many optimists would be quick to point out, problems can also be opportunities. While Soviet space enterprises were highly controlled segments of a strictly hierarchical military program, the space enterprises of the Newly Independent States (NIS) found themselves not only able to, but, in many cases, needing to establish contacts with their counterparts in other nations. Simply stated, for what was often the first time, former Soviet space concerns had to consider the bottom line, and they often looked to the West to make ends meet. Warmer East-West relations facilitated increased government-to-government collaboration in space, and this collaboration in turn allowed, and, in some cases, encouraged, commercial cooperation. U.S.-NIS commercial space cooperation is by its nature decentralized and therefore more difficult to analyze than governmental interaction, but, nonetheless, by the mid-1990s it was becoming an increasingly important part of the increasingly international and commercial space world.

Understanding the Russian Space Industry

Rocket Space Corporation (RSC) Energia

Before taking a closer look at specific U.S.-NIS commercial space ventures, it is useful to examine an example of a Russian space concern. An especially illustrative and germane example is RSC Energia. This case study includes some information which appears in other areas of the book.

RSC Energia has a proud history. The space enterprise was established in 1974 as the successor to the design bureau founded by Sergei Korolev, the father of Russian rocketry. It remains Russia's largest and most important space enterprise.

RSC Energia is located in Kaliningrad near Moscow, and has facilities in Samara and Primorsky. RSC Energia is Russia's main entity in charge of piloted spacecraft and heavy lift vehicles; it also operates Russia's manned space flight control center and employs all civilian cosmonauts. RSC Energia designed, led and, for ongoing programs, continues to direct the production of the Vostok, Soyuz, N-1 (lunar), Molniya and Energia launch vehicles, the Buran space shuttle, the Voskhod, Kosmos, Spektr and Progress spacecraft, the Salyut and Mir space stations, as well as several direct broadcast satellites. Since at least 1988 RSC Energia has also been developing an extensive satellite-based direct broadcast television platform; these efforts include the Signal, Yamal and Globis (initiated in 1990 as Marathon) projects. In 1991 RSC Energia employed roughly 35,000 people; by June 1994 employment had declined to approximately 30,000 people, and by July 1995 to about 23,000 people.

Just as RSC Energia is at the forefront of Russia's space program, so too is the enterprise at the vanguard of U.S.-Russian cooperation in space. NASA signed its first contract with a Russian firm with RSC Energia in Washington in June 1992 while Presidents Bush and Yeltsin held a summit meeting. That contract was to evaluate the use of the Soyuz spacecraft as a crew escape module for space station Freedom. This relationship expanded rapidly: in October 1992 Rockwell, the prime NASA contractor for the space shuttle, signed an $18 million contract with RSC Energia. Under this arrangement (which made RSC Energia a subcontractor on Rockwell's contracts with NASA), RSC Energia agreed to provide the docking interface mechanism and related support materials and services for space shuttle Atlantis dockings with the Mir space station.

Since then this cooperative relationship has expanded dramatically, and what was once a U.S.-led collaborative efforts with the Europeans and Japanese is now a U.S. and Russian-led space station Alpha (Chapter Five, pages 103-7). RSC Energia is the prime Russian contractor to NASA for space station Alpha, with responsibility for providing several space station modules, docking interfaces, life support systems, power generators and crew escape modules. Space station construction is scheduled to begin in late 1997, and in the meantime RSC Energia is leading U.S.-Russian cooperation in the Shuttle-Mir program. This cooperation is an integral precursor to more complex space station cooperation, provides the U.S. with an opportunity to learn from Russia's more advanced experience in long-term human stays in space, and provides the Russian space program with a much-needed infusion of capital (NASA is paying approximately $400 million for the astronaut time on Mir; Chapter Five, page 107).

The space station collaboration, which is officially government-to-government (NASA to the Russian Space Agency), helped lay the foundation for corporate cooperation. In the fall of 1994 Boeing and RSC Energia conducted a crystal experiment on Mir. Then, in August 1995 Boeing and RSC Energia reached an agreement to provide a service to facilitate low-cost placing of scientific, space development and remote-sensing hardware outside the Mir space station. These experiments were designed to spend several months in space, a great advantage over similar shuttle experiments, which featured about a twelve-day duration.

In order to expand its international business, RSC Energia established a permanent office, NPO Energia Limited in Virginia. RSC Energia received part of a $2.7-million dollar contract with the Department of Energy to conduct a study of power systems for large communications satellites (below, page 172). Energia also signed a contract with Daimler-Benz Aerospace (formerly Deutsche Aerospace) to study advanced space missions, such as lunar and environmental missions, and a separate contract with Rockwell and Daimler-Benz to investigate the commercial feasibility of RSC Energia's free-flying space servicing vehicle, the Inspector. Under the latter venture RSC Energia planned to launch an Inspector to Mir in January 1997.

In December 1992 RSC Energia joined Lockheed-Khrunichev-Energia International, which markets the Proton launch vehicle for commercial launches. RSC Energia produces the rocket's fourth stage. Commercial Proton launches cost about $70 million (compared to $70-120 million on the European Ariane launch vehicle), and LKEI has booked about $1 billion worth of launches through 2000. The first of its launches is scheduled for early 1996 (below, page 165).

RSC Energia is also one of four partners in the Sea Launch Limited Partnership, along with Boeing Commercial Space Company; NPO Yuzhnoe, Ukraine's premier aerospace enterprise and manufacturer of the Zenit launch vehicle; and Kvaerner, the Norwegian company which is Europe's largest ship builder. Sea Launch was founded in order to launch satellites from a mobile marine-based platform in the Pacific. RSC Energia planned to contribute integration and the upper stage to the venture, and the first launch was, as of mid-1995, scheduled for 1998.

NPO (Scientific Production Organization) Energia was one of the first large Russian military-related enterprises to seek permission to privatize. In March 1993 Yeltsin granted it permission to create a holding company, and in March 1994 he authorized the sale of stock. After corporatization NPO Energia was renamed Rocket Space Corporation (RSC) Energia, and

Yuri Semenov, the General Director, became the corporation's President. As per the latter decree, Energia privatized according to the Russia's "Type-1" plan. Twenty-five percent of the company's shares were distributed as non-voting stock to workers and retirees free of charge. Thirty-eight percent of the shares remained in the government's hands (this thirty-eight percent actually amounts to a controlling interest, since it is fifty-one percent of the voting stock). Nominally the government intended to hold this stock for three years, although it remains to be seen if the state will give up part or all of its stake. Of the remaining thirty-seven percent, twelve were sold in a closed auction in October 1994, ten percent were distributed to employees, five percent were distributed to management and ten percent were held in reserve to raise additional capital in the future. Management pushed hard for privatization as a means of raising capital, and used the invitation-only closed auction in order to control outside influence.

RSC Energia is central to Russia's spacecraft, launch vehicle, piloted exploration and unpiloted exploration programs. In fact, the space giant often challenges the authority of the Russian Space Agency. The Russian Space Agency is officially dominant: it is the closest Russian parallel to NASA. However, unlike NASA, the Russian Space Agency employs only a few hundred people. This makes the Russian Space Agency highly dependent on RSC Energia. Indeed, in key space policy meetings with top government officials, including Yeltsin and/or Chernomyrdin, Yuri Koptev (the Space Agency Director) *and* Semenov attend.

International Commercial Ventures in Space:
An Analysis by Sector

Although no company or venture falls into neatly drawn lines, it is useful to analyze international commercial space ventures by categorizing them by sub-sector (as in the following chart). Imagine you would like to watch television programs broadcast directly into your home via satellite. At the beginning, or left end of the space value chain, you would need a company to produce the power systems, satellite engines, batteries, antennae, flywheels and other subsystems necessary for the satellite. Each move to the right on the chain corresponds to an increase in value, culminating on the far right with a finished product. After the subsystems were completed you would need to contract out to electronics firms which could integrate the power systems with the circuits, instruments, etc., needed to receive, store and transmit signals. Then you would need a satellite integrator to build a satellite bus and integrate the satellite and all its systems into a launch vehicle. Now, in order to use your satellite you need a booster, but, before you can get that far along on the value chain your booster will need propulsion systems, which are often manufactured by companies other than the main booster producer. After you contract out the booster engines you acquire the booster, and then you need the support facilities to integrate the two and launch them into the desired orbit. Lastly, now that your satellite is orbiting, you need some one to maintain and operate the satellite.

Figure 1 provides an overview of a number of U.S. and NIS space concerns which fall into the categories, or sub-sectors, described above. Lines connecting companies represent collaborative ventures; when the companies involved have formed an actual joint venture, that company is represented as a box between the U.S. and NIS sections (e.g., Sea Launch). The diagram is representative of space collaboration in the first half of the 1990s, but is by no means complete. In some cases for clarity the diagram does not show all of the sub-sectors in which a given company operates. By showing a number of companies and the relationships they formed, we can gain a broader understanding of how the private sector approached post-Cold War space cooperation.[1]

U.S.-NIS Commercial Space Ventures: Dissecting the Diagram

Each of the collaborative ventures shown in Figure 1, along with a few others, are briefly described in this section. In the process many of the main U.S. and former Soviet space concerns are introduced. The organization of these summaries roughly follows the value chain in reverse, starting with launch vehicles and services, and then describing collaborative ventures in rocket propulsion; satellites; nuclear propulsion, satellite thrusters, other subsystems and spinoff technologies; and, lastly, space services ("operations and maintenance").

Lockheed-Khrunichev-Energia International

One of the most promising areas for U.S.-NIS commercial space ventures was launch services. For years the Soviet Union had placed more satellites into space than any other nation. Soviet boosters were known for their serial production, rapid launch capability and reliability. Meanwhile the market for commercial satellites, especially communications satellites, began to burgeon in the early 1990s.

The Lockheed-Khrunichev-Energia International (LKEI) joint venture was formed in January 1993 as the first U.S.-NIS joint venture to take advantage of former Soviet launch vehicle assets.[2] Lockheed Missiles and Space Company offered the ability to market the Protons in the West and to facilitate meeting Western requirements for satellite integration and technology transfer, Khrunichev manufactured the rocket and its first three stages and Energia produced the fourth stage (Energia profile,

U.S. - NIS Commercial Space Ventures

Boeing, Lockheed Martin

Subsystems

Atlantic Rsch. Corp.
Ball
Integrated Systems
Loral
Olin Aerospace
Orbital Systems
Pratt & Whitney
Unisys

Orbita

NPO Automatics and Instruments
OKB Faktel
NPO Geophysika
NPO Hartron
NPO Kompozit
Kurchatov (Topaz)
Kvant
NPP Nauka
TsSKB
NPP VNIIEM
NPP Zveda

Electronics

Aerospace Corn Dev (Canada)
Honeywell
ITT
Rockwell

Mashinostroenie

Satellite Integrators

Earthwatch
Final Analysis
Hughes
Loral
Motorola
Rockwell
TRW

SovCan

NPO PM (Applied Mechanics)

Makeyev
Mashinostroenie

Lavochkin

Propulsion

Alliant
Thiokol
Pratt & Whitney

McDonnell Douglas
Aerojet

Energomesh
Energia
Chexrmiantomatics
Dzu Dvoi Bimean
KB Lyulka
R&DIME
NPO Iskra
Salyut
NII Thermal Processes - KB
Chem Engineering

Launch Vehicles and Craft

Boeing, BER
Lockheed-Martin
Orbital Systems
Rockwell
Pack Astro

Energia
Khrunichev
Polyot
Samara
Yuzhnoe

Launch Facilities

NASA
USAF
Boeing
Lockheed Martin

Sea Launch   LKEI

Khrunichev
KB Transport Machinery
Polyot
NII KhIMMash

RSA, Military Space Forces, Strategic Rocket Forces

Operations and Maintenance

NASA, USAF
DOE
Allied Signal
Comsat, EG&G,
General Electric
Harris Corp.
Hughes, NEC,
Northrup Grumman
Orbital Systems
Qualcomm

ILS

Intersputnik
Molniya
KB Transport Engineering
TsNIIMash
NPO TP

Mt. Maidanak----?

Energia

U.S.

NIS

pages 159-62). In March 1995 Lockheed and Martin Marietta merged to form Lockheed Martin, and in June 1995 the new American space giant and Khrunichev formed a new joint venture, International Launch Services (ILS). ILS intended to market launches on the Lockheed Martin Atlas and Titan (heavy-launch) boosters along with the Proton. Due to the fact that the Proton and Atlas boosters have similar payload capacities, ILS planned to be able to offer superior launch flexibility (in terms of launch site and timing). At the time of the joint venture the European launch consortium Arianespace held about 50-60% of the commercial launch market, and ILS intended to take over the Europeans' dominant market share. "Together we will be the industry superpower," their President predicted.[3] Khrunichev was hoping to replace Energia's fourth stage with its own unit by 1998, thereby ousting Energia from the partnership. This Russian rivalry was an old one: Energia originated as Korolev's design bureau, and the Salyut design bureau (which Yeltsin made a part of Khrunichev; Chapter Six, page 134) descended from the space concern headed by Chelomei, one of Korolev's chief competitors.

The LKEI and ILS joint ventures were both centered around the Proton. By mid-1995 more than 200 Protons had been launched, and the rocket had accumulated a 96% success record. One version of the Proton (the four-stage SL-12) can place about 9,000 pounds in geosynchronous orbit, another (the three-stage SL-13) approximately 45,000 pounds into low-Earth-orbit. This impressive record, combined with competitive prices, gave customers considerable interest. By mid-1995 LKEI had booked launches through 2000 with PanAmSat, Inmarsat, the European Societe Europeenne des Satellites, Loral and Hughes. The total value of these launch contracts was about $1 billion, and the first launch was set for early 1996. The U.S. Overseas Private Investment Corporation (OPIC) agreed to provide $175 million in insurance for the first launch.

International ventures marketing the Proton had two general types of problems. First, the Proton depended on its deteriorating launch site, Baikonur (Chapter Seven, pages 147-52). Baikonur had four Proton launch pads, two of which were active when LKEI was formed. LKEI upgraded Baikonur facilities, adding clean rooms, work stations, fueling equipment and communications systems which met Western standards, as well as basic support infrastructure, such as a hotel, drinkable water and electrical systems. In mid-1995 ILS officials announced that they were scouting for another, equatorial launch site, although it was not clear if this was financially feasible.[4]

Second, there were political barriers. U.S.-Russian agreements limited the joint venture's bidding prices and total number of contracts, and American technology transfer protocol made their operations more cumbersome. These political problems will be discussed in more detail later in the this chapter.

Sea Launch

Sea Launch is a joint venture which brings together the largest Russian and Ukrainian aerospace companies, Energia and Yuzhnoe, with one of America's largest aerospace companies, The Boeing Company, and Europe's largest ship manufacturer, Kvaerner, to launch satellites on Zenit rockets based at a mobile marine platform. The companies began discussing the venture in early 1993; in June 1994 Boeing began formally seeking the necessary U.S. government licensing; and in May 1995 representatives of each of the four companies signed agreements formally establishing the joint venture. Sea Launch's headquarters was in Oslo, with offices in Moscow and Seattle (Boeing's headquarters).

Sea Launch's innovative plan was to convert an oil rig into a launch platform. In a typical sequence Yuzhnoe would build the rocket and Energia the upper stage; these would then be encapsulated in a Boeing-manufactured storage container and shipped to California, where the booster and satellite would be integrated and tested. The integrated vehicle would be placed on the converted oil rig and towed to sea by a command ship (both built by Kvaerner). This system was possible because the Zenit, which was first launched in 1985, was designed to be stored horizontally until right before launch. The actual launch site would be varied depending upon the type of orbit desired.

The Sea Launch venture would be able to place payloads weighing up to 11,000 pounds into geostationary orbit and up to 35,000 pounds into low-Earth orbit, and hoped to make its first launch in early 1998. Sea Launch was, like LKEI, forced to deal with U.S.-imposed launch limits (see below).

Cosmos Booster

On January 23, 1995 a Polyot-built Cosmos booster became the first Russian expendable launch vehicle to launch an American commercial satellite with a U.S. Commerce Department license (a Russian and Swedish satellite were also launched on the same rocket). Cosmos is an aerospace company based in Omsk, Siberia, that employed over 20,000 people at the time of the launch. In addition to the Cosmos light launch vehicle, Polyot has historically manufactured aircraft, strategic missiles, rocket engine subsystems and scientific, navigation and communication satellites. In fact, the Cosmos booster was derived from a medium range ballistic missile. The launch itself was managed by Russia's Military Space Forces. The satellite launched, FAISAT 1, was a store-and-forward communications satellite built by Final Analysis of Greenbelt, Maryland. FAISAT 1 was the first of what Final Analysis planned to be a 26-member constellation, and Polyot officials hoped that the American company would choose to launch all of its satellites on the Cosmos booster. Like other members of the former Soviet Union's military-industrial complex, Polyot was eagerly seeking conversion opportunities in order to maintain its operations.

This successful Cosmos mission, combined with the dearth of alternative small U.S. boosters, gave NASA an incentive to consider using Polyot's booster to launch some of its scientific payloads. NASA's scientific launch schedule was thrown into disarray by mid-1995 after the Pegasus booster (built by Orbital Sciences) failed twice. There were, however, significant bureaucratic hurdles to overcome. Federal regulations placed tight restrictions on launching government payloads on foreign boosters; no doubt these regulations would prove especially troublesome when the booster in question was Russian. Still, it was a remarkable sign of the times that NASA was even publicly considering such an option.

Converting ICBMs

Russia also began marketing converted SS-25 ICBMs for commercial launches. The SS-25 is a mobile intercontinental ballistic missile with a range of about 6,000 nautical miles; the Start-1 booster was designed to be able to place approximately 1,000 pounds into polar orbit. Because the launch vehicle is a converted ICBM, it falls under the jurisdiction of Russia's Strategic Rocket Forces, and not the Military Space Forces, like most other boosters. In 1994 the Moscow-based STC Complex, a private concern organized to market the converted ICBM, gained contracts to launch South African and Swedish payloads, but in March 1995 the booster exploded while attempting to orbit Israeli, Mexican and Russian satellites. The explosion was a setback, especially because it was Start-1's first attempt to launch commercial satellites (another version of the booster had been successfully tested with a Russian satellite in 1993). Despite this setback, it was quite likely that STC Complex, which was able to offer prices of roughly one-fourth its American competitors EER Systems and Orbital Sciences, would find other customers.[5]

Although Start-1 offered low prices, it also raised national security concerns. Representative Curt Weldon (R - Pennsylvania), the Chairman of the House Research and Development Subcommittee of the National Security Committee, voiced his concerns over using converted SS-25s as launch vehicles:

It [Start-1] is being marketed by a profit-making company in Russia, so it is no longer a part of the military arsenal. They could offer it to the Brazilians, which we told them we do not think would sit well, and supposedly they have plans to establish testing sites in South Africa and in other countries around the world.

But if you take the SS-25 system and you sell that--which eventually is going to happen--to an Iraqi or to a Libyan or to the Iranian government or perhaps even China or North Korea, they can put a chemical or biological weapon on that rocket and they have a missile capable of hitting America.[6]

In another effort to merge the interests of defense conversion and relatively inexpensive access to space, scientists at Stanford University, Lavochkin, TsNIIMash and Energia proposed using converted SS-18 missiles to launch scientific payloads, including robotic missions to Mars. Their grass root efforts, like those of the Start-1 proponents, faced an uphill battle.

Germany's Daimler Benz Aerospace and Russia's Khrunichev formally joined forces in 1995 in yet another venture to use converted ICBMs as space vehicles. In this case the joint venture, Eurorockot, sought to turn SS-19s into Rokot boosters. Rokot was intended to launch small satellites into low-Earth orbits, and was based from Plesetsk cosmodrome in northern Russia. Under the venture's plans, Germany's Ministry of Economics would forgive a portion of Russia's debt to Germany on a per-launch basis. However, like the Start-1 booster, Rokot suffered a setback when it exploded in a failed December 1994 launch attempt. Despite this failure the joint venture proceeded with plans for future launches. Even other former Soviet rockets were candidates for international commercial ventures.

A Rough Market for Entrepreneurs

There were many opportunities for utilizing effective, low-cost NIS boosters for international space ventures, but establishing and actually implementing the programs was a difficult process which often required a large American company willing and able to weather setbacks (difficulties are discussed in more detail below on pages 174-81). This fact is born out by two apparently failed ventures. Two different groups of American entrepreneurs, Commercial Space Management and Sea Launch Investors (not the Boeing-Khrunichev-Energia-Kvaerner venture) attempted to market Zenit and submarine-launched ballistic missiles, respectively as commercial boosters. Apparently neither venture was able to secure access to the Russian boosters.[7]

Upgrading American Boosters

In September 1994 McDonnell Douglas, the manufacturer of the Delta 2 launch vehicle, Delta Clipper Experimental reusable rocket, international space station hardware, missiles, commercial aircraft and electronics systems, signed a memorandum of agreement with Samara's Central Specialized Design Bureau and Progress factory. The two organizations planned to cooperate in launch vehicle activities, including rocket stages, system components and ground support systems.

Martin Marietta was considering using Energomash or Samara-produced engines to upgrade its Atlas booster. Energomash and Samara engines were being marketed by the American firms Pratt & Whitney and Aerojet, respectively (more below).

Finally, in late 1994 Martin Marietta began working with engineers from the Moscow-based Design Bureau of Transport Machinery (KBTM) to evaluate Russian Zenit launch processes. The Zenit benefits from a highly automated, efficient hanger-to booster-to orbit system, and Martin Marietta was interested in whether it could apply some of the Russian technology to its Atlas boosters. KBTM employed roughly 750 people, and did approximately 80% of its work for the Russian Space Agency and Military Space Forces, about 10% for the Russian navy, roughly 8% for conversion products (i.e., former military production lines producing consumer goods) and about 2% for foreign customers. In early 1995 Lockheed Martin renewed the contract with KBTM.

Pratt & Whitney, Lockheed Martin and Energomash

Pratt & Whitney is the space unit of United Technologies Corporation.[8] It builds space suits and fuel cells, but its main product is rocket engines. When the Iron Curtain fell Pratt & Whitney looked for opportunities to take advantage of the Soviet Union's advanced engine technologies. It found a partner in Energomash, one of Russia's leading producers of rocket engines. Energomash is a state-owned enterprise near Moscow that was founded in 1929. Its engines power the Zenit, Proton, Energia and Soyuz boosters.. Energomash had, by mid-1995, built more than 11,000 rocket engines which had flown on more than 2,300 launches. Many of the engines built by Energomash are able to operate under greater pressures and higher temperatures than their American counterparts, thereby providing more thrust.

Pratt & Whitney signed a Joint Marketing and Technology Licensing Agreement with Energomash in the fall of 1992, and as their relationship evolved the two companies formed a series of equity joint ventures. This agreement gave Pratt & Whitney the right to market Energomash's engines in the U.S., and exclusive rights to market the single-chamber RD-120 engine. Pratt & Whitney hoped to upgrade the RD-120 for the X-

34 reusable launch vehicle being developed by Orbital Sciences under NASA contract; they also hoped the engine could be used on the PA-2 booster being developed by the American company Pack Astro. In July 1995 Pratt & Whitney received delivery of a long-awaited RD-120 engine for extensive testing. The American company also hoped to use RD-700 series engines (derivatives of the RD-170) on the NASA Single Stage to Orbit (SSTO) vehicle under development.

Pratt & Whitney also hoped to use the two-chamber RD-180 engine for the booster on the U.S. Air Force's new Evolved Expendable Launch Vehicle (EELV) program. The USAF was planning to spend up to $2 billion between 1995 and 2000 to develop the medium- to heavy-lift booster. In order to be eligible for the American military program, the Pratt & Whitney had to be able to independently manufacture the engines in the United States within four years of the program's initiation. Pratt & Whitney was trying to negotiate an agreement with Energomash under which it would pay the Russian enterprise a licensing fee.

Lockheed Martin signed an agreement with Energomash in March 1994 to explore using the RD-180 engine to upgrade the Atlas booster. Pratt & Whitney and Rocketdyne (a division of Rockwell) were competing to manage the potential Atlas upgrades.

Aerojet and CADB, Lyulka, Samara and R&DIME

Aerojet formed several collaborative ventures with Russian propulsion concerns. One was with the Chemiautomatics Design Development Bureau (CADB) to adapt the RD-0120 booster for use on future American launch vehicles. The RD-0120 was originally built for the Energia heavy-launch vehicle; Aerojet hoped it could be used for an American Single Stage to Orbit vehicle.

Aerojet also worked with the Lyulka Engine Design Bureau to adapt the D-57 engine for use as an upgrade to McDonnell Douglas' Delta launchers and/or as a booster for SSTO testing.

Another Aerojet venture competed directly with Pratt & Whitney's attempts to adapt the RD-180 for use on the Air Force's EELV. The Samara State Scientific and Production Enterprise built the NK-33 engines for the Soviet Union's N-1 lunar rocket, but after the USSR canceled its efforts to land men on the moon, the enterprise was left with some seventy engines in storage. (Samara also produced the Soyuz, Progress and Molniya boosters.) Aerojet was helping to market these engines to Lockheed Martin, McDonnell Douglas and other Western firms. Like Pratt & Whitney, Aerojet hoped to be able to produce new NK-33s in the United States; it received shipment of an NK-33 for U.S.-based testing in July 1995. The Aerojet and Pratt & Whitney collaborative ventures were competing against the wholly American Rocketdyne effort.

Although it did not see a foreign market for Russian low-thrust engines, Aerojet felt it could gain valuable experience by testing the small rocket engines built by Russia's Research and Development Institute of Mechanical Engineering (R&DIME). In May 1994 the American company conducted extensive tests of R&DIME's LTRE 400N thruster (the 400N indicates that its thrust is 400 Newtons, a thrust level useful for attitude control, stabilization and orbit correction for satellites, space stations and other spacecraft). Aerojet officials claimed that the joint testing helped them learn what they had to do in order to gain U.S. government approval to import Russian rocket technology and engineers for collaboration; R&DIME officials hoped, nonetheless, that it might provide them with an opportunity to penetrate foreign markets.[9]

Satellites

In May 1994 Loral announced that it was negotiating a joint venture with Russia's large Mashinostroenie enterprise. Mashinostroenie was traditionally a primarily military enterprise. The discussions revolved around producing low-cost communications satellites.

One experience which most Western companies trying to work with their counterparts in the NIS had in common was that doing business with former Soviet space concerns was a new and different challenge that required longer than usual. This was certainly the experience of the Canadian firm Com Dev Limited, which formed a joint venture, SovCan Star, with NPO Applied Mechanics in May 1990. The consortium was founded to launch Applied Mechanics-manufactured telecommunications satellites specially designed to link East and West. The venture originally planned to launch the first of five satellites in 1996, but by early 1995 none of the satellites' construction had begun. As Garth Lewis, SovCan Star's Canadian program manager put it, "I know people will look at how long this has dragged on and conclude that we are no longer viable, but has taken more time than we thought...."[10]

Nuclear Power Systems, Satellite Thrusters, Other Subsystems and Spinoff Technologies

The United States stopped developing complete space nuclear power systems in the 1960s, but the Soviet Union continued advanced research in space reactors through the end of its existence. Nuclear reactors hold the potential of offering a relatively large amount of power for long periods of time, and, unlike solar panels, provide a consistent amount of power independent of their distance from the sun. This last characteristic is especially useful for interplanetary missions. Because space reactors are useful for early warning and anti-ballistic missile satellites, the Ballistic Missile Defense Organization (BMDO, the program office of the "Star Wars" program) began negotiating with the Russians in the government and at the Kurchatov Institute to study the advanced Soviet Topaz reactor. In May 1992 the University of New Mexico, on behalf of the BMDO, received shipment of two Topaz 2 reactors, and in March 1994 the university received four more. The USAF Phillips Laboratory, Los Alamos National Laboratory and Sandia National Laboratory worked with the University of New Mexico to study and test the Russian reactor. Although Star Wars funding had been severely cut, this was a relatively inexpensive way to catch up on decades of research. In late 1995 the joint Topaz program was to be transferred to the Defense Nuclear Agency.

In a separate collaboration, the Department of Energy gave Energia a $2.7 million contract to study power systems for large communication satellites. Under the contract, which Martin Marietta (and then Lockheed Martin) managed, Energia was to propose designs to provide 20-100 kilowatts of power for seven to ten years. Rockwell and General Atomics also participated in the study.

In late 1994 the small Silver Spring, Maryland-based engineering firm Orbital Systems announced that it had formed a joint venture with Ukraine's large space enterprise, NPO Hartron. Their company, Orbita, signed a contract with the BMDO to evaluate guidance, control and checkout techniques that allowed Soviet spacecraft to operate autonomously in orbit for extended periods. Hartron had long been a major supplier of computers, software and control systems for ballistic and cruise missiles, as well as for launch vehicles and satellites; the electronics company also produced air traffic control systems. The joint venture offered the U.S. the hope of gaining advanced Ukrainian technology, and gave the financially ailing Hartron a source of hard currency. The collaboration also held the opportunity to foster greater direct U.S.-Ukrainian cooperation in space.

In addition, Loral started an equity joint venture with the Experimental Design Bureau (OKB) Fakel in May 1992. Since 1955 Fakel had specialized in designing, testing and building electric and ionic satellite thrusters as well as ion and plasma generators. Soon thereafter the Loral-Fakel venture was named International Space Technology, Incorporated (ISTI), and the two concerns were joined by France's Societe Europeenne de Propulsion and the Research Institute of Applied Mechanics and Electrodynamics (RIAME) of the Moscow Aviation Institute. Fakel's plasma thrusters continued to be tested at NASA's Jet Propulsion Laboratories in mid-1995.

Olin Aerospace Company, a subsidiary of Olin Corporation that manufactures small satellite rocket engines, announced in June 1994 that it was in the midst of serious discussions with a number of Russian enterprises (more specific information was not made available to the public). Olin hoped to create a joint venture to co-develop electric propulsion systems.

In late 1994 Harris Computer Systems and the Moscow-based Lukon Financial Industrial Corporation announced that they would jointly develop and market computers for Russia's aerospace and defense industries.

In 1992 McDonnell Douglas and the Russian Academy of Science's Mechanical Engineering Research Institute (Imash) agreed to cooperate on a series of space technology research projects. The collaboration centered on materials, advanced mathematics and extended human stays in space.

A group of Russian scientists and American business executives gave conversion an interesting twist in 1992 when they formed a joint venture, Scifor, to develop extremely large television displays. Scifor's displays were designed for sports arenas, concert halls, shopping malls, etc., and boasted that they were several times "thinner" than their competitors. The original source of the technology: the Soviet space program.[11]

Representatives from the United States Air Force, along with the Department of Energy, Los Alamos National Laboratory, Idaho National Engineering Laboratory and Phillips Laboratory visited Russia in the summer of 1993 hoping to learn from Russia's advanced optics and laser technologies. These technologies are applicable to, among other things, tracking and vaporizing space debris. The American delegation visited the Vavilov State Optical Institute, the Institute of Applied Physics, Astrofizika (an entity which developed anti-missile defense systems), the Russian Institute of Space Device Engineering and Kometa (a space concern which specialized in anti-missile and reconnaissance systems). At one point each of these organization's very existence was secret, and many of their ongoing activities remain classified. During their 1993 visit the American team awarded a $20,000 contract to Passat, a private Nizhny Novgorod-company to conduct feasibility studies of a system to track small objects in low orbits. However, in early 1994 the Americans were forced to cancel a return visit. Their Russian counterparts had expressed a "heightened sensitivity" over their proposed collaborative projects' dual-use (military and civil) applicability.[12] Their heightened sensitivity was most likely related to the fact that Russia had, by 1994, begun to pursue less pro-Western policies (some termed this shift "Russia's Reverse;" Conclusion, pages 189-90). To paraphrase comments President Yeltsin made in May 1995, "Russian scholars should cooperate with Western research institutions but only with a clear understanding of the need to defend Russia's national interests."[13]

Space Services

Lastly, some American and former Soviet space concerns joined teams to provide complete satellites services. For example, General Electric American Communications (GE Americom) announced in early 1995 that it would beam English- and Hindu-language television programming to the India subcontinent via Intersputnik's Express 6 satellite. Intersputnik was the Soviet Union's closest parallel to America's Comsat Corporation.

In another Mashinostroenie-involved venture, Sokol-Almaz-Radar (SAR, a spinoff from Mashinostroenie) formed a venture with the Washington-based Sokol Group. The two enterprises hoped to find investors to fund the Almaz 1B satellite for commercial remote sensing operations. As of mid-1995 the venture was still seeking funding.

In 1993 the U.S. Air Force Space Command began investigating cooperation with Uzbekistan's Mount Maidanak observatory to contribute to the Air Force's ability to track space debris. The formerly classified remote observatory boasts a high-resolution optical telescope which incorporates technologies not used in the West. One of the highlights of Mount Maidanak's construction is a large circular deck which girdles the telescope; the deck improves telescope performance by limiting temperature fluctuations. In 1995 the Air Force allocated its Phillips Laboratory $100,000 to explore collaboration with the Uzbeks involving Mount Maidanak and other unique technologies, including some of the advanced laser and optics technologies described above.[14]

Finally, in yet another sign that the Cold War was indeed over, three American companies, Central Trading Systems, Lambda Tech International and Aerial Images, signed a contract with the Russian concern Sovinformsputnik to sell remote sensing images. The images were to be collected by Russian military satellites.[15]

The Difficulties Facing Post-Cold War
U.S.-NIS Commercial Space Ventures

Although it took intergovernmental space cooperation to allow and sometimes encourage nongovernmental collaboration, as Yeltsin's comment cautioning Russian researches illustrates, political factors continued to limit cooperative ventures in the post-Cold War world. Just as American proliferation concerns threatened intergovernmental collaboration (Chapter Seven, page 152-3), Russians feared selling national security assets to an all-too-new ally. On the other hand, as the extensive Topaz project demonstrates, these dual-use concerns could be overcome. There were two other challenges to U.S.-NIS commercial space ventures. First, there was the fact that the American and (former) Soviet cultures were extremely different and had spent the majority of the century not only developing in relative isolation from each other, but also developing largely in opposition to each other. American and NIS business and space executives had to overcome their very different business practices in order to succeed. Second, there were other political barriers to commercial cooperation in space, especially American limits on NIS launches of Western commercial satellites.

The Business of Cultural Differences

There were several challenges to American firms hoping to conduct commercial space ventures with the NIS that were unique to doing business in that part of the world. The most obvious example of these challenges is the fact that these ventures had to overcome several governmental barriers on both sides. In addition to space applications' military relevance and launch limits (below), the U.S.-Russian space ventures had to overcome other aspects of governmental regulations. Customs were a common problem. For example, Russians had a hard time getting the Topaz reactors out of the country, and the American side barely managed to be able to receive and then deliver the reactors. The first shipment of Topaz reactors was enabled only by a last-minute phone call by Vice President Quayle. In another (albeit non-commercial), powerful example of the problems of customs, it took personal intervention by Vice President Gore and Prime Minister Chernomyrdin to overcome a delay caused by Russian customs; the 754 kilograms of equipment delayed was needed for American astronaut Norman Thagard to use on board the Mir space station.[16]

American businesses also faced problems in areas they were normally able to take for granted. For instance, transferring money to Russians was problematic, partly because of Russia's underdeveloped financial infrastructure, and partly because it was difficult to get the money into the hands of the people for which it was intended. In addition, Russia's tax and duty system was complex and unstable. There were other challenges involved with communication, including those intrinsic to doing business with people who speak another language (such as translations, etc.), but also those due to the fact that Russia had, by American standards, a poorly developed communications infrastructure.

Another challenge American businesses hoping to conduct business with the Russians faced is that Americans and Russians seemed to view negotiations differently. While Americans tend to assume that once something is agreed upon it is final, Russians see negotiations as a process, with agreements subject not only to renegotiation but even to unilateral withdrawal.[17] An especially difficult topic for U.S.-Russian negotiations was intellectual property rights (IPR), both because IPR is critical for a high-technology field such as space, and because Americans and Russians (where IPR was not as strong a concept as in the United States) had very different idea of what amounted to honoring intellectual property agreements.

Finally, American firms were often frustrated by the limited information they could find on the Russian industry. When an American space executive became interested in a Russian (let alone Uzbek) space enterprise, he could not simply pull up a page full of data from his Dow Jones server. What Americans regard as basic data was difficult if not impossible to acquire about NIS enterprises. This made evaluating the market, availability of products and services and actual capabilities extremely difficult. It also often forced the American partners to trust their counterparts without the option of verifying their claims. This monopoly of information proved especially problematic when one partner in a venture with more than one NIS participant attempted to serve as the sole conduit of information with the American partner(s).

Launch Limits

There is perhaps no better introduction to the topic of launch limits than to introduce a Russian point of view:

> Russia is rightfully placing its hopes in the space business....It has not yet lost the status of a leading space power and it has the necessary scientific potential and production capacities. We can carry out up to 40 commercial launches a year, and this is half of the world capacities [sic]. This gives rise to various kinds of predictions about the future. Optimistic, bold, and in a certain sense even extremely concrete....It is not so good when these prophecies do not coincide with the real course of events. The international space technology market has been in existence for more than 15 years and the competition is just as stiff here as it is in other international markets. Today the...ratio looks approximately like this: 60-64 percent is the share of Western Europe, 30-35 percent--the United States, and the share that goes to Russia, the largest space power, is ridiculously small: [a]ccording to various estimates it is from 0.5 to 3 percent. The United States has ended up with the powerful lever for regulating this market, and it does not want to let go of it.
> What about the program for cooperation in space between Russia and the United States?[18]

What about the U.S.-Russian program for cooperation in space? While it is true that the United States and Russia worked together on an intergovernmental and commercial basis when it was to all parties' mutual benefit, the launch market was largely another story. Commercial launches were more about business and protectionism, i.e., competition, than about cooperation.

A Promising Market

Commercial boosters generate revenues by launching satellites, and the satellite market was expanding rapidly in the post-Cold War world. An early 1995 industry survey counted that there were 949 satellites planned for launch between 1995 and 2004.[19] The majority (68%) of these missions were commercial, and most of the planned launches were of communications satellites. One planned venture held the potential to almost double the number of satellites launched during that decade: Teledesic, a wireless communication system backed by Microsoft founder Bill Gates and McCaw Cellular's Craig McCaw intended to build an 840-satellite network. The U.S. Department of Transportation's Space Transportation Office calculated that satellite communications was already a $6.5 billion annual industry by 1995. Before 1995 that sector of the market had grown by at least ten percent each of the past several years, and analysts expected the growth rate to increase in the late 1990s and early twenty-first century.[20]

An increasing number of launch vehicles jockeyed for these launches. The Arianespace Ariane 4 and Ariane 5, Lockheed Martin Atlas, McDonnell Douglas Delta, Russian Proton, Chinese Long March and Japanese H-2 all hoped to carve out a share of the market to launch satellites weighing 1.8 metric tons or more. The newly developed Lockheed Launch Vehicle, Orbital Sciences' Pegasus and Taurus launch vehicles, Ukrainian Yuzhnoe's Zenit, Russian Polyot's Cosmos and other boosters were competing for the launch of smaller satellites. Rebrov's estimates for market share were only somewhat similar to estimates made by the *New York Times,* which calculated that the European consortium Arianespace held 55% of the market, the U.S. Atlas and Delta launchers slightly less than 30% and the Russian Proton and Chinese Long March the rest.[21]

The great majority of these one thousand or so satellites would be built by American manufacturers. Seven of the world's ten biggest satellite manufacturers in the first half of the 1990s were American, including Lockheed, Martin Marietta, Hughes, Loral and TRW. Of the nearly one thousand, by far the most would be built by Lockheed Martin, but NPO PM (Applied Mechanics) was the second largest manufacturer.[22]

By the time the Iron Curtain fell, American space companies were already more than frustrated that they had lost most of the commercial launch market over which they had once held a monopoly. Arianespace, the European consortium founded in the early 1970s, had, by the late 1989s, largely realized its founding purpose: to dominate the commercial launch market. While most American commercial boosters were converted ballistic missiles, the Ariane launch vehicles were, from the beginning, intended to launch satellites. Thus, when former Soviet enterprises began trying to enter the commercial market, some American launch companies, such as Martin Marietta, General Dynamics, McDonnell Douglas and Orbital Sciences, appealed to the U.S. government to protect their interests. As Michael Wash, president of General Dynamics' Commercial Launch Services expressed his (somewhat exaggerated) fears, "Proton has the capability to put us, Ariane and everybody else out of business if there are not constraints put on them."[23] Other American companies, such as Lockheed and Boeing, lobbied the U.S. government to allow their cooperative ventures to launch commercial satellites.

Unfortunately for the Boeing's, the Lockheed's, the Russians and the Ukrainians, the protectionists won the first round. The U.S. government was able to control NIS entry into the commercial launch market largely because U.S. firms dominate the world's satellite manufacturing capability, and the American government must provide all the appropriate licenses for U.S.-built satellites to be launched on foreign rockets. The U.S. used this lever, along with the promise of Shuttle-Mir and space station cooperation, to resolve one of its major concerns about Russia and Ukraine: proliferation. When Vice President Gore and Prime Minister Chernomyrdin agreed in September 1993 to merge the two countries' space station efforts, they also signed a bilateral agreement whereby Russian promised to adhere to the Missile Technology Control Regime (MTCR; Chapter Five, page 108). It also wielded this power to conclude another agreement whereby Russia was allowed to launch up to nine contracts to launch Western commercial satellites into geostationary orbit through 2000 (eight plus one satellite contract concluded before September 1993), and up to three contracts to put satellites into low-Earth orbit. (The limits applied to launch contracts, not the number of satellites; thus, one launch could include several satellites). Also under this accord, which was originally described in positive terms such as "commercial entry" but was later portrayed as a "launch limit" agreement, Russia agreed not to offer geostationary launch prices more than 7.5% below their Western competitors'. Technically this limit was placed upon Russia to protect American companies from state-subsidized competitors in "non-market" economies.

The entry of former Soviet boosters into the commercial launch market was highly charged. The fighting among launch firms was made even more complicated by the interests of satellite manufacturers and end-users, who stood to gain from the less expensive access to space that Russian boosters offered. The U.S. Shuttle-Mir and space station programs were dependent upon the maintenance of the NIS's space infrastructure, and commercial launches could easily provide the NIS's space program with more money than the American government. In another complicating factor, some forces within NASA and the Air Force opposed granting Russia greater access to commercial space because they felt it would devalue the X-33, X-34 (NASA) and Evolved Expendable Launch Vehicle (USAF) boosters, which were under development. Even these NASA and Air Force interests were complicated by the fact that certain parts of those same organizations favored using Russian engines to make their boosters more affordable, and by the fact that some of their contractors were advocating this application of Russian engine technology. In addition, entry into the commercial launch business was a piece in the much larger chess game of U.S.-Russian relations, a gift parsed out to the Russians for other concessions (such as adherence to the MTCR). If the United States had granted Russia full access, it would have lost this lever for future proliferation and other concerns.

The debate became even more complicated in 1994. That March LKEI began complaining that it had already reached the eight-launch limit.[24] In September separate U.S. trade delegations traveled to Russia and China to renegotiate commercial launch agreements (the U.S. agreement with China was signed in 1989 and was set to expire at the end of 1994). The Russians watched the U.S.-Chinese negotiations very carefully to see if they were being treated on an equal basis.[25] In October Lockheed officials complained to the House space subcommittee that the Ariane launcher was more heavily subsidized by France than the Proton was by Russia, and that the Proton would be taking business away from the European launcher, not from American boosters.[26] That January the United States agreed to grant China eleven geostationary launches and the right to underbid by 15%.[27] It seemed that Russia interpreted this as treating the Chinese more favorably, and in December 1994 Prime Minister Chernomyrdin personally raised the issue with Vice President Al Gore during their semi-annual meeting.[28]

The issue continued to be a contentious one in 1995. That February 1995 a U.S. government official who had helped negotiate the agreement with the Chinese acknowledged that the United States had been conducting internal discussions about whether or not they should relax the limits on Russian launches. In May the Chairman and Chief Executive Officer of Lockheed Martin sent Vice President Gore a letter arguing that the United States had to eliminate the limits, and the following month, the President and Chief Executive Officer of McDonnell Douglas wrote the vice president a letter arguing the exact opposite, claiming, among other things, that this could undermine his company's development of the Delta 3 launch vehicle (not-so incidentally, the Delta 3 would compete directly with Lockheed Martin's Titan).[29] Rumors abounded that the limits would be relaxed or lifted. Meanwhile U.S.-

Russian relations stumbled over controversial Russian sales of technology to Iran, China and Brazil (Chapter Seven, page 190-91). These issues clouded Clinton's visit to Russia in May and the Chernomyrdin-Gore meeting in June/July. Nonetheless, in July the Director General of Khrunichev, Anatoly Kiselov, announced that the United States had agreed to eliminate the launch limits: "As diplomats say, we have found a mutual understanding.[30] Although officials publicly announced that Gore and (RKA Director) Yuri Koptev had discussed the issue, there was no (diplomatic) announcement of a new understanding.

Yet another problem loomed on the horizon. A little-publicized section of the September 1993 agreement stipulated that the Russians could not launch more than two geostationary missions within a twelve-month period. This constraint did not stop LKEI from scheduling three geostationary launches for 1996. Unless the agreement was modified or the third launch was cancelled or rescheduled, there would be a major problem.

As if all the issue of geostationary limits was not complicated enough, there were four issues revolving around launch quotas which made the situation even more complex. First, there was no blanket agreement constraining Russian bids on launches to low-Earth orbit. This market was so unpredictable (for example, the future of the 840-satellite Teledesic venture earlier mentioned was unclear in 1994/early 1995; this constellation alone would dramatically alter the market) that American officials could not outline any framework. Instead the two parties agreed to approve low-Earth launches on a case-by-case basis. Moreover, even the 7.5% constraint for geostationary bids seemed arbitrary: Frank Weaver, who helped negotiate Russia's entry into the commercial launch market, disclosed:

> In the Russian agreement we had no experience at the time that agreement was negotiated. During the consultations there was just a number that was arrived at because no one really knew what it would be, and that number just happened to be 7 1/2 percent.[31]

Second, there were more than two parties involved in many Russian launch vehicles. As of mid-1995 the United States still had not reached a similar agreement with Ukraine, even though Ukraine had, shortly after Leonid Kuchma became president in November 1994, signed on to the MTCR. This left commercial launches of Western satellites on the Zenit vague if not impossible. In July 1995 Ukraine and the U.S. resumed negotiations over a bilateral space launch trade accord, with the outcome uncertain.[32] Third, it was not clear how to define services that were truly international, such as those Sea Launch intended to provide. Was the launch service Ukrainian since the bulk of the booster was built in Ukraine, was it Russian because it built the main engines, or was it an American booster because it was integrated in the United States? Finally, it was questionable whether or not the agreements' percentages were effective, since no launch contract was attacked for underbidding, even though some company officials did accuse China and Russia of violating the pricing rules.[33]

In short, the launch limits were inadequate because they were overly simple rules which, whatever their merit, did not apply to a space business that was rapidly becoming truly international. The quotas were part of the bureaucracy's last gasping attempt to throw up barriers which could define "national space programs." If the world continued to evolve along the post-Cold War trends already clear by 1995, commercial ventures of the kind discussed in this chapter would continue to proliferate, and these artificial rules would give way to truly international space programs. If the market was allowed to have its way, the extremely competitive launch services sector, combined with satellite service providers' demand for inexpensive access to space, would eventually overcome the lobbying power of those who wished to maintain limits. However, if American relations with Russia and/or Ukraine worsened, commercial ventures could disappear even more quickly than intergovernmental cooperation.

Opportunity Breeds Collaboration

After the collapse of the Soviet Union, Russia may have been, by American standards, sorely lacking business acumen and overflowing with logistical difficulties, but the fact remained that it was also a nation brimming with advanced technology. For decades the Soviet Union had poured incredible amounts of assets into its space program, and this cultivation yielded fruits that American companies could not resist trying to harvest.

Although working with the Russians often proved frustrating, it also promised enormous returns on investment. It may, for example, have taken Pratt & Whitney almost three years to receive shipment of an Energomash RD-120 engine for testing, but developing a similar power plant from scratch would have cost approximately $1 billion.[34] Finally, the Russians and their counterparts in the rest of the NIS were eager to cooperate. They were working desperately to prevent an implosion of their country's space program: in 1994, for instance, Russia's space industry production fell 38.6%.[35] Finally, there were the people behind the TASS and NASA announcements, people who had long respected their distant competitors; now they could finally not only look behind the curtain but actually try to pull it back and join forces to conquer humankind's last frontier--together.

Notes

1. All of the information presented in this chapter was collected from nonproprietary sources.

2. The company was officially incorporated in March 1993 in Delaware.

3. Peter B. de Selding, "ILS Unites Atlas, Proton Sales," *Space News* 19-25 Jun. 1995: 9.

4. de Selding, "Unites," 9.

5. Ben Iannotta, "Russia Accused of Underbidding to Win Greensat Launch," *Space News* 28-6 Feb./Mar. 1994: 8.

6. Interview. *Space News* 29-4 May/Jun. 1995: 22.

7. Andrew Lawler, "East-West Launch Vehicle Venture in Turmoil," *Space News* 8-14 Aug. 1994: 5.   Jeffrey M. Lenorovitz, "U.S. Entrepreneurs Seek Russian SLBMs," *Aviation Week* 19 Apr. 1993: 22-23.

8. The other five units are Hamilton Standard aviation systems, Otis elevators, Carrier air conditioners, Sikorsky helicopters and United Technologies automotive systems.

9. Debra Polsky Werner, "Aerojet Fires Russian-Made Thruster Rocket in U.S. Tests," *Space News* 23-29 May 1994: 20.   Michael A. Dornheim, "Aerojet Tests Russian Oxidizer-Cooled Thruster," *Aviation Week* 27 Jun. 1994: 75.

10. Peter B. de Selding, "SovCan Star," *Space News* 13-19 Feb. 1995: 21.

11. "Scifor Corp.," *PR Newswire* 16 Jun. 1994.

12. William B. Scott, "Russian Politics May Stymie Laser/Optics Collaboration," *Aviation Week* 21 Mar. 1994: 49.

13. "Yeltsin Warns," *Monitor* vol. 1, no. 19, 25 May 1995.   Paraphrased from *Podmoskovskie Izvestia* 23 May 1995.

14. William B. Scott, "Uzbek Site," *Aviation Week* 15 May 1995: 68-69.

15. Doug Clarke, "U.S. Companies to Market Russian Military Satellite Images," *OMRI* no. 145, pt. 1, 27 Jul. 1995.

16. Chapter Five, page 104.   Ben Iannotta, "U.S., Russia Learn as They Go," *Space News* 21-4 Nov./Dec. 1994: 1.

17. Andrei Baev, Matthew J. Von Bencke, David Bernstein, Jeffrey Lehrer and Elaine Naugle, "American Ventures in Russia," (Stanford: Center for International Security and Arms Control, 1995) 20.

18. Mikhail Rebrov, "Mirages in the Visibility Zone, or In the Space Market Without Changes," *Krasnaya Zvezda*. As translated by FBIS.   "Space Services Market Called Unfair," *FBIS* 24. Feb 1995: 1-2.

19. James R. Asker, "Commercial Growth Key to Space Sector," *Aviation Week* 13 Mar. 1995: 97.

20. Asker, "Growth," 95.

21. Richard W. Stevenson, "Way Ahead in the Space Race," 5 Apr. 1995: D1. See also: Andrew Wilson, *Jane's Space Directory* 1994-95 (Coulsdon: Jane's Data Division, 1994) 7-10.   "Russia Staking Its Claim in Commercial Launch Market,"

Aerospace Daily 28 Dec. 1994: 425. Peter B. de Selding, "Five Launchers Competing for Full-Sized Payloads," *Space News* 8-14 May 1995: 8. Peter B. de Selding, "Launch Vehicle Firms Predict Market Shift," *Space News* 8-14 May 1995: 8. Peter B. de Selding, "Launch Market Prepares for a Business War," *Space News* 7-13 Mar. 1994: 8. James R. Asker, "Racing to Remake Space Economics," *Aviation Week* 3 Apr. 1995: 45-48, 53.

22. James R. Asker, "Nearly 1,000 Spacecraft," *Aviation Week* 16 Jan. 1995: 55.

23. General Dynamics manufactured the Titan. General Dynamics was bought by Martin Marietta, which then merged with Lockheed. Patrick Seitz, "U.S. Officials Probe Proton, Long March Pricing Policies," *Space News* 7-13 Mar. 1994: 12.

24. Debra Polsky Werner, "Proton Venture Nears Sellout on Agreement," *Space News* 21-27 Mar. 1994: 3.

25. Andrew Lawler, "U.S. To Begin Launch Talks With China, Russia," *Aviation Week* 12-18 Sep. 1994: 20.

26. Paul Mann, "Washington Outlook," *Aviation Week* 3 Oct. 1994: 25.

27. Craig Covault, "Russian Proton Challenges Ariane," *Aviation Week* 24 Apr. 1995: 40.

28. Warren Ferster, "Russia: Relax Launch Limits," *Space News* 19-25 Dec. 1994: 1, 21.

29. Warren Ferster, "U.S. Firms Lock Horns Over Launch Quotas," *Space News* 12-18 Jun. 1995: 1, 37.

30. "Russian Expert Says U.S. to Lift Satellite Quotas," *Reuters* 30 Jun. 1995. "U.S. Said to Life Satellite Restriction," *OMRI* no. 130, pt. 1, 6 Jul. 1995.

31. Mr. Weaver was director of the Department of Transportation's Office of Commercial Space Transportation. "Official Says Russian Launch Agreement Could Be Revisited," *Aerospace Daily* 14 Feb. 1995: 233.

32. "Ukraine, U.S. Resume Launch Accord Talks," *Space News* 10-16 Jul. 1995: 2.

33. Lawler, "Launch Talks," 20. Richard W. Stevenson, "Russian Rockets Finding Eager Customers in West," *New York Times* 17 May 1994: D21.

34. William Harwood, "Pratt & Whitney Touts RD-120 Rocket Engine," *Space News* 31-6 Jul./Aug. 1995: 8.

35. This can be compared with an average of 38.8% for the entire military-industrial complex that same year. "Signs of Stabilization in Defense Industries," *Interfax* (17 Jan. 1995) Translation by FBIS, *FBIS Daily Report--Soviet Union* 18 Jan. 1995: 40.

Conclusion:
The Space Age
Outlives the Cold War

Scanning the Age

Throughout the space age, U.S.-Soviet/Russian competition and cooperation in space have coexisted. It has never been an either/or proposition; the levels of competition and collaboration have varied over time.

This coexistence is a natural product of the fact that American and Soviet/Russian space policy has always been both a reflection and an instrument of domestic and foreign policy interests. In short, space policy is the confluence of a number of influences. When the confluence of interests motivating the space policies of both nations overlap, the nations collaborate; when it benefits both space powers to compete (recall, for example, Academician Sedov's especially candid 1960 remark that "If we really cooperated on man-in-space, neither country would have a program because the necessary large support in money and manpower was only because of the competitive element and for political reasons."),[1] they compete.

As the analysis of the varying influences upon the Eisenhower and Khrushchev leaderships shows, American and Soviet leaders have always faced a myriad of considerations in forming space policy (Chapter One). The American and Soviet governments are multi-branched entities which try to promote their best interests, as they judge them, as well as they can. Thus, Khrushchev decided to rely heavily on the space program to bolster the Soviet and Communist Party images and to strengthen his own position in power. In addition, Khrushchev used the related development of intercontinental ballistic missiles to reduce overall military expenditures and allocate greater resources to domestic concerns, primarily agriculture. As a result the early Soviet rocketry program was geared largely toward developing ICBMs and achieving impressive "firsts." So, too, was it a result of a collection of policy influences that the Eisenhower administration made the explicit decision not to race the Soviets into space, but instead to keep the civilian and military space programs as separate as possible, to publicly promote space for peace and to use the first civilian satellite as the vanguard of a military space program largely geared toward reconnaissance. It was only after the shock of the public relations success of Sputnik that NASA was created and the United States fully engaged itself in the image battle of the space age.

This public relations battle led to the mutual pursuit of space one-upmanship. From 1957 to 1969 the United States and the Soviet Union rapidly expanded their space efforts, driven largely by their desire to achieve more, more quickly and, most importantly, more than the other space power. Somewhat ironically, the public relations battle which was a part of this competition largely facilitated the first attempts at cooperation in space (Chapters Two and Three). These two chapters demonstrate the fact that cooperation in space must be understood as existing within the context of the nations' overall space programs and relations. In order to appear to be the superpower most interested in promoting other nations' welfare, the United States and Soviet Union competed in an effort to be recognized as the space power most willing explore and exploit space for the benefit of all mankind. In addition, the United States and the Soviet Union tried, largely in the framework of the United Nations, to promote that version of space law which would most benefit themselves; sometimes, as the world's only major space powers, these interests were similar. Thus, between 1957 and 1969 the U.S. and USSR managed to lay the foundations of space law and cooperation in space even as they raced to the moon.

Although Kennedy and Khrushchev personally exchanged suggestions of greater cooperation in space, meaningful U.S.-Soviet collaboration was not to come without an improvement in overall superpower relations. Détente marked the most significant Cold War improvement in U.S.-Soviet relations, and so too did the 1975 Apollo-Soyuz Test Project (Chapter Four) mark what was then the pinnacle of U.S.-Soviet cooperation in space. The Apollo-Soyuz mission marked a confluence of the United States' and the Soviet Union's interests in space: both superpowers did their best to appear to be putting aside national rivalries for mankind's sake, and both nations benefitted from their first closeup view of the other's space program. The joint mission also served the space powers' selfish interests: the mission served as a bridge between America's Apollo and Space Shuttle eras, and as a balm to the Soviet pride, which had been smartly wounded by the loss of the moon race.

Despite plans for ongoing collaboration, U.S.-Soviet relations worsened considerably in the late 1970s, and cooperation in space all but disappeared. Warm overall relations remained a precondition for collaboration in space. By the time détente had passed, improved technologies had expanded military applications of space, and, more than ever, space became a tool of the Cold War arms race.

Then, beginning in 1985, superpower relations began to improve dramatically, opening up the possibility of renewed cooperation. Furthermore, both nations' individual space programs' budget cuts increased American and Soviet incentives to reduce costs by working together. While the Russian civilian space program actively sought hard currency, President Clinton pursued a foreign policy success by linking U.S. support of Russian democratization to collaboration in space. These parallel developments led to unprecedented levels of cooperation in space (Chapter Five). In September 1993 the United States and Russia agreed to merge their space station plans, in February 1994 Sergei Krikalev became the first Soviet cosmonaut launched on an American spacecraft and in June 1995 the space shuttle *Atlantis* completed the first of seven Shuttle-Mir dockings. The improved relations and confluence of other (largely budgetary) interests facilitated greater cooperation than had ever occurred.

Still, the heightened level of cooperation of the late 1980s and early 1990s was not without difficulties and portents of trouble to come (Chapters Six and Seven). The end of the Cold War left Russia in a shambles economically, socially and politically and the former Soviet civil space and science communities struggling to maintain their budgets and personnel, to deal with organizational chaos and to cope with inherited and largely anachronistic conservatism. In addition, the dissolution of the Soviet Union left the former Soviet space program spread among several newly sovereign nations, and relations among the Newly Independent States proved problematic. Russian space woes were highlighted by troubles at the Baikonur Cosmodrome and delays of unpiloted missions to Mars. This sea of troubles threatened Russia's ability to serve as a valuable partner and forced American policy makers to begin reevaluating the degree to which they should depend on the Russians even as they worked toward fulfilling the collaborative agreements they had forged. Lastly, the pause in cooperation caused by the U.S.-Russian disagreement over technology exports to India, Iran and Brazil served as a stark reminder that some things had not changed: space was still unavoidably related to missile technologies, and the U.S. and Russia remained different powers with different interests.

Era of Cooperation?

Nonetheless, the early 1990s were a promising time for collaborative space efforts. The United States and Russia were engaged in unprecedented levels of cooperation. As President Clinton stated in his

State of the Union address on January 26, 1994,

> As we take these steps together to renew our strength at home, we cannot turn away from our obligation to renew our leadership abroad. This is a promising moment. Because of the agreements we have reached this year--last year--Russia's strategic nuclear missiles soon will no longer be pointed at the United States nor will we point ours at them. Instead of building weapons in space, Russian scientists will help us to build the international space station.[2]

The American and Russian space collaboration promised lower costs, broader applications and shared ideas and experience. Specifically, during the early 1990s the U.S. and Russia, due to their tightening budgets and the improvements in space technology, sought more and more to tailor their space programs toward greater direct benefits to mankind. Thus the early 1990s boded especially well for cooperation in environmental, remote sensing (such as resource detection and monitoring) and communications and data management applications. Even the Central Intelligence Agency Director, R. James Woolsey, promoted such cooperation: in a November 19, 1993 address, Mr. Woolsey said,

> earlier this year in my meetings with Mr. Primakov [the head of the Russian Intelligence Service]...I suggested to him that Russia and the United States could begin to help each other in tackling some environmental problems such as water pollution by swapping some photos. After all, going back many years, I have the best pictures of Lake Baikal and he has the best ones of the Great Lakes.[3]

In addition, manned space efforts are intrinsically more expensive than unpiloted projects, and, thus, the U.S. and Russia had a common interest in merging their manned missions (which were intended to culminate in space station Alpha). By the mid-1990s Russia's economy seemed that it might be on the road to recovery, but it had a long way to go. As long as Russia's economy languishes, the Russian civilian space program is likely to look to the West, and primarily to the United States, for a partnership in space.

Russia's Reverse?

In late 1993/early 1994 some Americans began wondering about "Russia's reverse," and President Clinton's administration felt compelled to show that it was not being too "soft" on the Russians.[4] As discussed in Chapter Six, in late 1993 reformists did poorly in the parliamentary elections, while the ultra-conservative Zhirinovsky gained a frightening amount of support. Then in early 1994 President Yeltsin raided his cabinet, eliminating several leading advocates of market and democratic reforms.

These developments alone would have caused American leaders to pause and reflect on the course of events in Russia. Two other events gave them further reason to take stock of the U.S.-Russian relationship. First, on February 17, 1994, Russia unilaterally announced that it had negotiated a settlement of the siege of Sarajevo, just in time to save the Serbs (who were besieging the Bosnian capital and had historically enjoyed Russia's patronage) from North Atlantic Treaty Organization (NATO) air strikes. This assertive diplomatic maneuver abruptly interrupted the NATO-led Western course of action, and, American officials stated, came as a surprise to the United States.[5] Second, on February 22, 1994, the Central Intelligence Agency announced that one of its chief counterintelligence agents had been serving the Soviet Union/Russia as a double agent for eight years. The United States angrily responded with diplomatic expulsions; the Russians returned the favor.[6] These spy wars resembled the Cold War more than a new alliance.

The United States was not the only party reconsidering U.S.-Russian relations in early 1994. Russian leaders were becoming increasingly bitter and wary of NATO's Partnership for Peace plan (Chapter Seven, page 145), and were growing increasingly impatient with Western dictates regarding their economy. As the Russian Foreign Minister Andrei Kozyrev wrote in a *New York Times* op-ed piece titled "Don't Threaten Us,"

> it appears that some Western politicians, in Washington and elsewhere, envision Russia not as an equal partner but as a junior partner. In this view a 'good Russian' is always a follower, never a leader....Russia is destined to be a great power, not a junior one.[7]

However, there had been no reversal. Instead these events served to correct the opinions of those who, in getting caught up with the euphoria of the end of the Cold War and the improvement in U.S.-Soviet/Russian relations, had forgotten that not even close allies always agree. In the mid-1990s the United States and Russia were neither intimate partners nor enemies. In addition to the fact that their friendly relations had only begun to develop some ten years before, the improved U.S.-Russian relations were tempered, and always will be, by the fact that the American and Russian governments represent two distinct collections of peoples with two distinct sets of interests.

Military Versus Civilian Space Programs, Post-Cold War

Since space policy is a reflection and instrument of domestic and foreign policy influences, and given that the United States and Russia will continue to have common and conflicting interests, the American and Russian space programs will continue to compete as well as cooperate. In the early 1990s this competition quietly took place in both nations' military space programs.

While this story is one that will best be told after more information is declassified, some things are, as of 1995, already clear. First, the United States military's reliance on space grew in the 1990s. Advanced technology, the 1991 Persian Gulf War against Iraq, increased dependence on rapid response tactics and expanding proliferation concerns combined to increase America's emphasis on military uses of space. The Persian Gulf War in particular demonstrated the effectiveness of military space systems; Lt. Gen. Thomas S. Moorman, then the commander of Air Force Space Command, noted that during this war, "for the first time, we have space beginning to become fully integrated into the prosecution of hostilities, to a much greater extent than...[before]."[8] Lt. Gen. Donald M. Lionetti, the commander of the Army Space and Strategic Defense Command, said that without the Global Positioning System the American ground forces' success in Iraq would have been impossible.[9] Second, reflecting this increased reliance on military space applications, the United States spent a growing percentage of its shrinking (until 1995) defense budget on space. In fiscal year 1986 the space investment budget was less than half of the total investment in the Army, whereas for fiscal year 1993 it exceeded investment in the Army by twenty percent.[10] Publicly known U.S. Department of Defense space spending exceeded NASA's budget every year from 1982 to 1995.[11] It is safe to assume that total U.S. military space spending, when taking into account classified expenditures, was significantly greater during this period.

Meanwhile, through the middle of the 1990s Russia managed, amazingly enough, to maintain its military space constellations at Cold War levels. Given the general disarray in Russia and Russian aerospace, Western defense analysts had predicted that the former Soviet military space program's capabilities would not be maintained.[12] However, beginning in late 1992 the Russians began maintaining and even expanding the Soviet military satellite constellations. This surge continued into the following year: during the first four months of 1993, the Russians launched twelve new military spacecraft, including three advanced navigation satellites, two electronic intelligence (ELINT) satellites, two imaging reconnaissance spacecraft, two missile warning satellites, two comsats and a new-generation ocean surveillance spacecraft.[13] For all of 1993 the Russians successfully launched thirty dedicated military missions, and in 1994 they had twenty-six successful dedicated military launches.[14] From January to early August 1995 Russia launched eight successful military missions. In addition to these "dedicated" (or fully) military satellites, several satellites launched each year, such as geodetic satellites (which measure the earth's gravitational and magnetic fields), had civilian and military applications.

In fact, not only had Russia managed to maintain the majority of its 1990 military constellations' capabilities: by 1995 it had in fact expanded some of its capacities. For example, although the number of ELINT 3 (the "3" refers to the generation) satellites in orbit had dropped from six in 1990 to three in 1995, this was a natural phase-out of the older spacecraft, which was being replaced by the ELINT 4 constellation. In 1995 the ELINT 4 constellation remained full at four spacecraft. Although Russia launched many fewer photographic reconnaissance satellites than the Soviet Union, this was primarily because Russia had all but stopped launching a kind of "photo bird" which the Soviets had launched as many as thirty-five of each year. Instead Russia introduced more advanced reconnaissance satellites which lasted longer and, in some cases, relayed observations in real or near-real time by digital communications (instead of dropping recoverable film packs, as the older common Soviet type did). In August 1994 Russia launched a new type of reconnaissance satellite (*Cosmos 2290*) and earlier that same month it launched an undefined new military satellite (*Cosmos 2285*). Russia launched fewer navigation, early warning and ocean surveillance satellites than the Soviet Union, but, again, this was because their newer satellites were lasting longer. As a result in 1995 two of the three navigation networks were maintained at four and six satellites, while another was slightly expanded to twenty-one; one early warning constellation was maintained at nine spacecraft while the geosynchronous launch detection network was expanded from three to four; and the two ocean surveillance constellations' populations remained steady at four and two. Moreover, Russia's two low-altitude military communications constellations were in fact improved from their 1990 status: one of the two constellations was maintained at three satellites, while the other one was phased out for another, newer twenty-four satellite network. The other two main categories of Russian military comsats, the Raduga and data relay satellites, were also maintained at 1990 levels.[15]

Another major sign of Russia's space program's post-Cold War health is that in 1994 the Russians launched six new types of satellites--a truly banner year. Three of these spacecraft (the *Gals 1* and *Ekspress 1* advanced comsats and the *Coronas 1* scientific satellites) were purely civil, one (the *Electro 1* geostationary weather satellite) had civil and military applications and two (the *Cosmos 2290* advanced reconnaissance and classified *Cosmos 2290*) were new military satellites. Another sign of the program's health is that, by 1995, piloted launches and launches supporting human space exploration seemed to have stabilized at seven to eight launches annually.

Given all of the problems facing the former Russia and its scientific and aerospace communities, the Russians' successful expansion of the Soviet military space satellite systems reflects what must have been a remarkable prioritization of military space objectives. Indeed, some estimated that in late 1993 approximately three-fourths of Russia's research and development funding was military.[16] These Russian efforts were apparently a continuation of those managed under the Soviet Union in the late 1980s, i.e., also under less-than-optimal economic conditions. Roald Sagdeev, Director of the Institute of Space Research from 1973 to 1988, estimated that the Soviet space program was 85-90% military.[17] About 70% of Soviet space launches in the late 1980s were military.[18]

There is other evidence that space was a high priority for Russia's military. Russian Ministry of Defense statistics state that Russia's Ministry of Defense spent only 34% less on procurement of space satellites and space launch vehicles in 1992 than the Soviet Union did in 1991. This drop is small compared with procurement in other key military areas:

System Type	Reduction in Value of Procurement (%)
Intercontinental Ballistic Missiles	55
Submarine Launched Ballistic Missiles	39
Tactical Missiles	81
Surface-to-Air Missiles	80
Air-to-Air Missiles	80
Aircraft	80
Tanks	97
Field Artillery	97
Multiple Rocket Launchers	76[19]

The New "New Look"

The fact that the United States and Russia were, in the early 1990s, placing ever higher priorities in military applications of space is reminiscent of Eisenhower and Khrushchev's New Look policies of the 1950s.[20] Both Eisenhower and Khrushchev sought to gain more defense with less money, and both turned to technology. The frontier of technology in the 1950s was nuclear weapons; in the early 1990s it was space technology. Whatever the future of competition and cooperation in space, military space systems will likely continue to play a vital if not dominant role in determining the United States' and Russia's overall space policy.

That does not necessarily mean that competition will dominate U.S.-Russian space interaction. If cooperation in space and in general progressed into the late 1990s and beyond, it is conceivable that the American and Russian military space programs would begin working together, at least to some extent. In a potential harbinger of such a development, Jeffrey Harris, Assistant Secretary of the U.S. Air Force and the head of the National Reconnaissance Organization, traveled to Russia in July 1995 to "explore how the two nations' intelligence satellites might be used to provide 'derivative products' to environmental satellites."[21] Although such interaction falls more into the kind that CIA Director Woolsey mentioned (above), it may well have been the beginning of true U.S.-Russian *military* space cooperation.

The Future of Space Policy

So what does the future hold for U.S.-Russian cooperation in space? The best crystal ball we have is history. One thing is clear: increased collaboration in space will continue to be dependent on warm overall relations. Ongoing cooperation in space is dependent upon the stabilization--economically and politically--of a pro-Western Russian government. Through mid-1995 this stabilization proved elusive. The emergence of an anti-West Russian regime would, like the end of détente in the late 1970s, quickly bring an end to cooperation in space. Nonetheless, no matter what happened, U.S.-Russian space interaction would never be the same after the close cooperation of the early 1990s. Just as the Apollo-Soyuz mission paved the way for future collaboration, the American and Russian cooperation of the early 1990s had set a new precedent, a new standard.

Lastly, the two great space powers have never cooperated in space merely for the sake of cooperation itself; instead, cooperative space efforts have flourished only when they are facilitated by warm overall relations *and* compatible goals in space. In other words, cooperation has flourished only when it has proven competitive, i.e., only when it has presented both the United States and the Soviet Union/Russia with advantages over competition. The early 1990s saw a greater agreement among U.S.-Russian interests in space than ever before, and thus these years were marked by unprecedented cooperation in space. Since the dawn of the space age, many had hoped that this new frontier, which was beyond national boundaries, might prove to be the grounds for transnational cooperation and an usher of greater peace on Earth; in many ways their hopes seemed to be coming true in the post-Cold War world.

Despite this surge in joint work in space, the United States and Russia will continue to have differing interests, in general and in space. The end of the Cold War did not change some of the fundamental characteristics of states and their relations with each other. Neither the American nor the Russian government is monolithic; while NASA and the Russian civilian space program benefitted from increased cooperation in the early 1990s, American and Russian military planners resisted yielding and in fact worked to expand their Cold War-level space programs. Thus, organizations within both nations will continue to have an impact. So too do top-level leaders in both governments have an effect on space policy. For example, President Clinton's decision to pursue a joint U.S.-Russian space station gave cooperation in space significant impetus, but, had he decided that the United States could not afford to help develop a space station, events would have turned out quite differently. Still, individuals' influence is constrained by the larger currents of the nations' overall space programs and relations, as Kennedy and Khrushchev's largely failed proposals for greater cooperation evidence (Chapter Two, pages 48-50 and 51-55).

By the middle of the 1990s one of the greatest hopes for the future of U.S.-NIS space cooperation was that the private sector had begun establishing extensive joint commercial space ventures (Chapter Eight). These commercial ventures offered the potential to draw the United States and the countries of the Newly Independent States closer together, both in general by making their economies more interdependent, and more specifically in their space efforts. The private sectors could some day prove a buffer to the stormy times in U.S.-NIS relations, constituting interest groups which lobby their governments to maintain and encourage commercial space cooperation. This was especially important in the post-Cold War world, when the commercial space market was becoming more important and the former Soviet Union's private sector was a growing influence on government policy.[22]

Indeed, space policy, with its multifaceted coexistence of competition and cooperation, is a reflection of the complicated nature of government. U.S. and Soviet/Russian space policy offers an excellent paradigm for a broader understanding of U.S. and Soviet/Russian governments and U.S.-Soviet/Russian relations. The space age outlived the Cold War, but space policy remained subject to the same fundamental rules of domestic and foreign considerations which operated during that era of superpower competition.

Notes

1. Chapter Two, page 50.
2. "Excerpts From President Clinton's Message on the State of the Union," *New York Times* 26 Jan. 1994: A16.
3. R. James Woolsey, "The Future of Intelligence on the Global Frontier," Address to the Executive Club of Chicago, 19 Nov. 1993: 9.
4. "Learning From Russia's Reverse," editorial, *New York Times* 29 Jan. 1994: A14. Steven Greenhouse, "U.S. to Russia: A Tougher Tone and a Shifting Glance," *New York Times* 21 Mar. 1994: A9. Steven Greenhouse, "Christopher Defends Russia Policy in Wake of CIA Agent's Arrest," *New York Times* 3 Mar. 1994: A6.
5. John Kifner, "Serbs Withdraw in a Russian Plan to Avert Bombing," *New York Times* 18 Feb. 1994: A1. Elaine Sciolino, "Move by Russians a Surprise to U.S.," *New York Times* 18 Feb. 1994: A6. Two weeks later, on March 1, 1994, American NATO planes shot down four Serbian aircraft that had violated the "no-fly" zone. This was the alliance's first combat action in its 45-year history.
6. David Johnston, "Ex-Branch Leader of CIA is Charged as a Russian Agent," *New York Times* 23 Feb. 1994: A1. Steven Erlanger, "Amnesty of Foes Brings Disarray to Yeltsin Team," *New York Times* 28 Feb. 1994: A1.
7. 18 Mar. 1994: A29.
8. Robert L. Butterworth, *Guide to Space Issues for the 1990s* (Los Alamos: Center for National Security Studies, 1992) 3-4.
9. James R. Asker, "Space Key to U.S. Defense," *Aviation Week* 3 May 1993: 57.
10. U.S. Senate, Committee on Armed Services, *National Defense Authorization Act for Fiscal Year 1993 Report* 102nd Congress, 2d Session Rpt. 103-352 (Washington: GPO, 1992) 85.
11. National Aeronautics and Space Administration, *Aeronautics and Space Report of the President, Fiscal Year 1992 Activities* (Washington: National Aeronautics and Space Administration, 1993) 96. Ben Iannotta, "U.S. Spy Satellite Spending Static in 1995." *Space News* 14-20 Feb. 1994: 4.
12. Craig Covault, "Russian Military Space Program Maintains Aggressive Pace," *Aviation Week* 3 May 1993: 61.
13. Covault "Aggressive Pace," 61.
14. The 1993 dedicated military launches included seven photographic reconnaissance missions, three ELINT satellites, four ELINT Ocean Reconnaissance Satellites (EORSATs), three early warning satellites, three low-altitude navigation spacecraft, one launch of three GLONASS positioning satellites, one radar calibration satellite, two Raduga comsats, three Molniya comsats and three low-altitude photographic reconnaissance missions (including two "six-pack" configurations).
The 1994 dedicated military launches included seven photographic reconnaissance missions, two ELINT satellites, one EORSAT, two early warning satellites, one low-altitude navigation spacecraft, three GLONASS positioning missions (each mission carried three satellites), one radar calibration satellite, three Raduga comsats, one Molniya comsat, four other military comsats, two "six-pack" low-altitude photographic reconnaissance missions and one new military

satellite with an unknown function.

15. Nicholas L. Johnson, telephone interview, 24 Mar. 1994, and personal interview, 18 Aug. 1995. Mr. Johnson is an expert in foreign space programs and is the author of *The Soviet Year in Space,* yearly editions from 1981 to 1990, as well as *Europe and Asia in Space,* which covers 1991-1992. Mr. Johnson is currently a Senior Scientist at Kaman Sciences Corporation.

16. Peter Aldhous, "Can Russia Slim Down to Survive?" *Science* 19 Nov. 1993: 1201.

17. Leonard David, "Boosters versus Bread Lines," *Ad Astra* Jul./Aug. 1992: 6.

18. Craig Covault, "Russia/CIS Space Outlook Chaotic But Critical to Global Planning," *Aviation Week* 16 Mar. 1992: 125.

19. Stockholm International Peace Research Institute, *SIPRI Yearbook* 1994 (Oxford: Oxford UP, 1994) 458-59.

20. Chapter One, pages 10-11 and 12-15.

21. Paul Mann, "Who'da Thunk It?" *Aviation Week* 26 Jun. 1995: 19.

22. Matthew Evangelista's *Technical Innovation and the Arms Race* (Ithaca: Cornell UP, 1988) provides a useful context for analyzing the potential importance of Russia's private sector on Russia's space policy. While in the Soviet Union technical innovation usually started from the top of the system and flowed down, by the middle of the 1990s in Russia this trend had begun to reverse, and lower echelons of society (e.g., individual enterprises) had more opportunities to innovate and influence.

Appendix A

US GDP in Constant Dollars

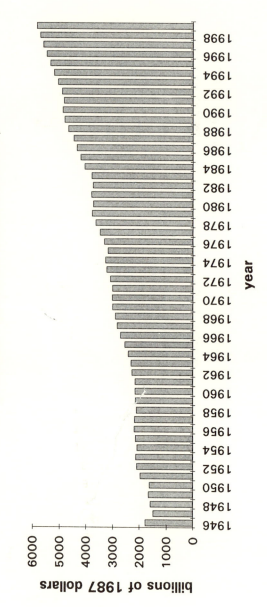

Source: *Budget of the United States Government, Historical Tables, Fiscal Year 1996* (Washington: GPO, 1995). Figures for 1996-1999 are estimates. Transition Quarter (July-September 1976) not shown in this or any other graph.

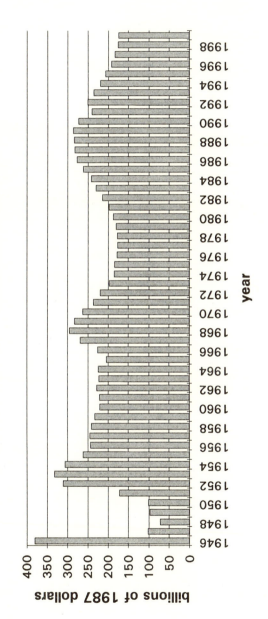

U.S. Defense Spending in Constant Dollars

Source: *Budget of the United States Government, Historical Tables, Fiscal Year 1996* (Washington: GPO, 1995). Figures for 1996–1999 are estimates.

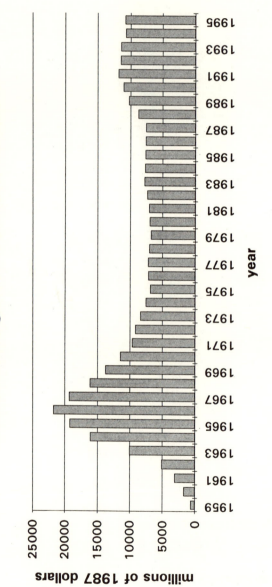

NASA Budget in Constant Dollars

millions of 1987 dollars

year

Source: *Budget of the United States Fiscal Year 1996* (Washington: GPO, 1995).

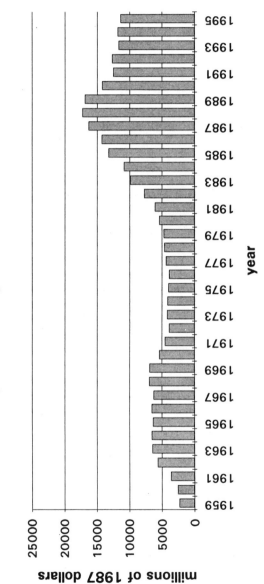

DOD Space Budget in Constant Dollars

Sources: 1959-1991: *Aeronautics and Space Report of the President: Fiscal Year 1992 Activities* (Washington: NASA, 1993). 1992-1995: Ben Iannotta, "U.S. Spy Satellite Spending Static in 1995," *Space News* 14-20 Feb. 1994: 4. Figure for 1995 is an estimate. These numbers serve only to indicate general trends in U.S. military space spending, as they do not include classified budgets, such as the space expenditures of the CIA, National Security Agency and National Reconnaissance Office (whose very existence remained classified until September 18, 1992).

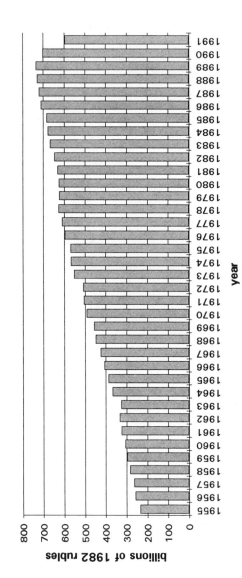

USSR GNP in Constant Rubles

Sources: 1955-1987: U.S. Congress, Joint Economic Committee (JEC), *Measures of Soviet GNP in 1982 Prices*, S. Rpt. 101-128 (Washington: GPO, 1991) 54-57. 1988 and 1989: JEC, *Allocation of Resources in the Soviet Union and China*, Hearing April 20, May 16 and June 28, 1990, S. Hrg. 101-476, Pt. 15 (Washington: GPO, 1991) 30. 1990: JEC, *Global Economic and Technical Change*, Hearing May 16 and June 28, 1991, S. Hrg. 102-586 (Washington: GPO, 1991) 152. 1991: JEC, *Global Economic and Technological Change*, Hearing June 8 and July 27, 1992, S. Hrg. 102-586, Pt. 2 (Washington: GPO, 1992) 94. All figures are CIA estimates. For explanation of CIA methodology, see *Measures*. For discussion of the accuracy of CIA and other estimates, see Lev Dudkin and Anatol Vasilevsky, "The Soviet Military Burden: A Critical Analysis of Current Research," *Hitotsubashi Journal of Economics* vol. 28, 1987: 41-61.

USSR Defense Spending in Constant Rubles

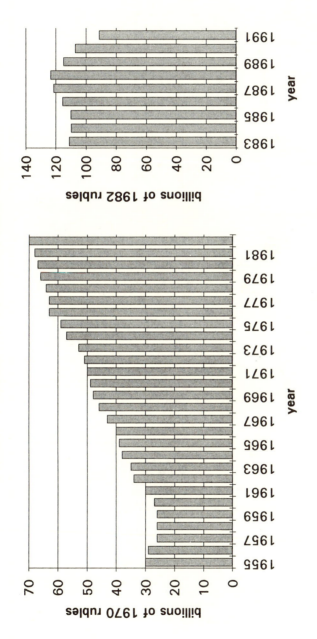

Sources: 1955-1982: Dudkin and Vasilevsky: 43. Figures are based on a compilation of CIA estimates. 1983-1991: JEC, *Global* Pt. 2: 13, 36.

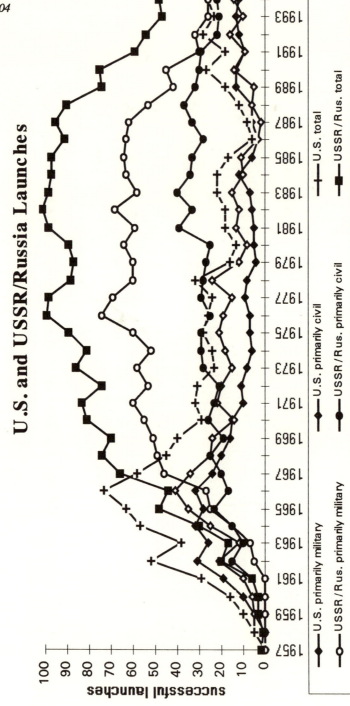

U.S. and USSR/Russia Launches

Legend:
- U.S. primarily military
- USSR/Rus. primarily military
- U.S. primarily civil
- USSR/Rus. primarily civil
- U.S. total
- USSR/Rus. total

Y-axis: successful launches

Successful launches are defined as those attaining earth orbit or beyond. Enumerates launches, not spacecraft (some launches orbited multiple spacecraft). Source: Marcia Smith, "Space Activities of the U.S., C.I.S. and Other Launching Countries/Organizations," *Congressional Research Service Report for Congress* 95-873 SPR, 31 Jul. 1995.

Appendix B:
Chronology of the Space Age

1945

February 4-11	Yalta meeting: Roosevelt, Stalin and Churchill sign the Declaration on Liberated Europe
May 8	V-E Day ends European theater of World War II
June 5	Germany divided into four occupation zones by U.S., USSR, France and UK
June 26	San Francisco Conference ends; UN Charter is agreed upon
July 27-August 1	Potsdam meeting: Big Three finalize European borders and the division of Germany
August 6	U.S. drops atomic bomb on Hiroshima
August 9	U.S. drops atomic bomb on Nagasaki
August 14	V-J Day ends Pacific theater of World War II

1946

	USSR violates Tehran agreement to withdraw from Iran as agreed, but eventually pulls out under Iranian protests and UN pressure
March 6	Churchill makes "Iron Curtain" speech in Fulton, Missouri
May	Civil war breaks out in northern Greece between monarchists and communists
May 26	Czech communists win 38% of the vote and set up a one-party government
September 2	U.S. and 18 Latin American countries sign the Rio Treaty on collective security

1947

	Political turmoil in Italy: increasingly assertive and violent Communist minority causes the U.S. to increase its aid and to delay the withdrawal of troops on November 29
March 12	Truman announces the Truman Doctrine policy of opposing the spread of communism
June 5	Secretary of State Marshall calls for massive U.S. aid to Europe; the "Marshall Plan" is passed on March 31, 1948, with initial grants and loans of $17 billion to 16 nations

July 25   National Security Act passes, creating the U.S. Air Force
                   (USAF), Central Intelligence Agency (CIA), Joint Chiefs of Staff
                   (JCS), National Security Council (NSC) and Department of
                   Defense (DoD)
October 14   Chuck Yeager breaks the speed barrier, flying a USAF X-1

1948   USSR and Yugoslavia relations worsen as Tito asserts his own
                   communist style
May 15   Israel formed; Egypt, Jordan and Syria attack the next day; war
                   ends July 20, 1949
June 24-   First Berlin crisis: USSR blockades Berlin, U.S. responds with
   May 12, 1949   airlift

1949   Chinese Communists gain control of mainland China;
                   Nationalists flee to Formosa
March 20   USSR cuts off talks on a unified German currency
April 4   NATO formed by U.S., Great Britain, France, Belgium, Canada,
                   Denmark, Holland, Iceland, Italy, Luxembourg, Norway and
                   Portugal; Greece and Turkey join February 1952;
                   Germany joins May 1955; Spain joins May 1982
May 15   Hungarian communists form a government after one-party
                   elections
August 29   Soviets test their first nuclear bomb

1950
January   USSR withdraws from the Atomic Energy Commission
January 31   Truman approves hydrogen bomb R&D
February 14   USSR and China sign agreement of mutual assistance; USSR
                   returns Manchurian railroad
March   NSC-68 released
March 23   UN forms the World Meteorological Organization
June 2   North Korea invades South Korea

1951
April 11   Truman dismisses General Douglas MacArthur
September 1   U.S., Australia and New Zealand sign the Pacific Security
                   Treaty, forming ANZUS
September 8   U.S., Japan and 47 other nations sign Treaty of Peace with
                   Japan, restoring its sovereignty

1952
May 27   Belgium, France, Holland, Italy, Luxembourg and W. Germany
                   sign a mutual defense pact
October 3   Great Britain explodes its first atomic bomb
November 1   U.S. successfully explodes the first hydrogen bomb, a non-
                   deliverable device named "MIKE" at Eniwetok Atoll (10 MT)

1953

March 5	Stalin dies
June 16	Workers riot in over three hundred places in East Germany
July 27	Armistice concludes Korean War
August 12	Soviets explode Joe-4, which they claim is the world's first deliverable H-bomb (smaller yield than MIKE)
October 30	NSC concludes report 162/2, introducing Eisenhower's "New Look" policy

1954

January 12	Secretary of State Dulles announces "massive retaliation" nuclear policy
March 1	U.S. explodes BRAVO, 15 MT deliverable hydrogen bomb at Bikini Atoll
May 7	French decisively defeated in Vietnam at battle of Dien Bien Phu; July 27 Vietnam divided
May 17	U.S. Supreme Court declares "separate but equal" unconstitutional in "Brown vs Board of Education of Topeka, Kansas," starting desegregation of schools
June 18	CIA-supported Guatemalan exiles invade Guatemala, form a government July 8
September 8	U.S., Great Britain, France, Australia, Cambodia, Laos, New Zealand, Pakistan, Philippines, South Vietnam and Thailand form Southeast Asia Treaty Organization (SEATO)

1955

January 18	First Quemoy and Matsu (Formosa straits) crisis: Communist China seizes a Nationalist island, continues shelling; U.S. announces intention to defend
March 16	Eisenhower announces that the U.S. would use nuclear weapons in the event of war
April 15	USSR signs treaty with Austria, agreeing to pull its troops out (U.S., Great Britain and France sign on May 15)
May 6	West Germany formally joins NATO
May 13	Warsaw Pact formed
May 22	Khrushchev and Bulganin visit Tito in Belgrade, beginning reconciliation with Yugoslavia
July 18-23	Geneva Conference of Big Four; Eisenhower's "Open Skies" proposal (July 21)
July 29	U.S. announces intention to launch a satellite as a part of the International Geophysical Year (IGY)
July 30	USSR announces intention to launch a satellite as part of IGY

1956

February 24	Khrushchev denounces Stalin at 20th CPSU Congress in "secret speech"
June	U-2, product of Lockheed's Skunk Works, begins flying over

	USSR
July 26	Nasser nationalizes Suez Canal
October 19	Khrushchev, Kaganovich, Mikoyan and Molotov go to Warsaw to protest the election of Gomulka as the Polish Communist Party's First Secretary
October 29	Israel invades Egypt, the Suez Canal crisis lasts until November 7
November 4	USSR attacks Hungary, decisively suppressing Hungarian Revolution

1957

January 5	Eisenhower announces the Eisenhower Doctrine policy of opposing the spread of communism in the Middle East
January 10	Eisenhower, in State of the Union, proposes mutual international control of "outer space missile and satellite development"
July 1	IGY begins (ends December 31, 1958)
August 8	U.S. Army Ballistic Missile Agency (ABMA) launches a Jupiter-C rocket; its nose cone is the first man-made object successfully ever recovered from space
August 9	U.S. joins the International Atomic Energy Agency, joining USSR and 79 other members
August 27	USSR announces possession of ICBM after successful tests
September 9	Eisenhower signs the Civil Rights Act of 1957 into law
October 4	USSR launches *Sputnik*, the first artificial satellite
November 3	USSR launches *Sputnik 2*, 1121 lbs., carrying the first animal, a dog ("Laika"), into space
November 7	U.S.'s Gaither Report recommends rapidly expanded missile programs, R&D and a national fallout shelter program; report leaked in December
November 14	UN adopts Resolution 1148 (XII), hoping to ensure that space is used exclusively for "peaceful and scientific purposes"
December 6	American *Vanguard TV-3* launch failure; "Kaputnik," "Stayputnik," "Flopnik"

1958

	Under Secretary of State for Political Affairs, Eugene Rostow, publishes *Stages of Economic Growth, A Non-Communist Manifesto,* advocating massive U.S. Third World aid
January 31	U.S. launches the ABMA's 10.5 lb *Explorer 1* with micrometeroid detectors, Geiger counter and telemetry equipment; discovers Van Allen radiation belts; first U.S. satellite orbited
February 1	Syria and Egypt form the United Arab Republic (UAR)
March 17	U.S. launches *Vanguard 1* launched with Geiger counter, proton-precession magnetometer; provides first space measurements of the Earth's shape and uses the first satellite solar cells

March 26	U.S. launches the ABMA's *Explorer 2*
May 14	USSR launches 1.5-ton *Sputnik 3*, the first geophysical laboratory
July 14	U.S. Marines land in Lebanon after pro-Western Iraq government falls; British troops land in Jordan
July 29	National Aeronautics and Space Act of 1958 enacted, creating NASA
August 6	DoD reorganized, President's Science Advisory Committee and post of Special Assistant to the President for Science and Technology created
August 18	NSC-5814/1, "Preliminary U.S. Policy on Outer Space" completed
late August	Communist China begins shelling Nationalist islands, starting the second Quemoy and Matsu crisis
September 2	National Defense Education Act passed
November 24	UN Ad Hoc Committee on the Peaceful Uses of Outer Space (COPUOS) founded
November 27	Khrushchev demands West abandon West Berlin, causing second Berlin crisis
December 18	U.S. launches *Score*, the first active communications satellite

1959	U.S. tests the first antisatellite (ASAT) weapon, an air-launched Bold Orion missile
January 1	Castro takes power in Cuba, deposing Batista
January 2	USSR launches *Luna* (Moon) *1*, the first spacecraft to escape the Earth's gravity
February 17	U.S. launches *Vanguard 2*, which takes the first space photograph of the Earth
March 3	U.S. launches *Pioneer 4*, which flies by the moon
March 14	U.S. submits first concrete proposal for space cooperation at Committee on Space Research (COSPAR) meeting
May 11	Geneva Foreign Ministers' meeting resolves Berlin crisis
August 7	U.S. launches *Explorer 6*, which broadcasts the first TV pictures from space
September 12	USSR launches *Luna 2*, first spacecraft to hit the moon
September 15-27	Khrushchev tours U.S., discussing "peaceful coexistence" with Eisenhower at Camp David
October 3	USSR launches *Luna 3*, takes the first photographs of the moon's dark side on October 7
December 1	U.S., USSR and 10 other nations sign Antarctica Treaty as a follow-up to IGY
December 12	UN Resolution 1472 (XIV), adopted unanimously, establishes permanent COPUOS

1960	
January 2	Eisenhower approved NSC-5918, "U.S. Policy on Outer Space"
February 13	France explodes its first atomic bomb

April 1	U.S. launches *TIROS 1*, the first weather satellite on Thor-Able rocket
April 13	U.S. launches navy's *Transit 1B*, the first successful navigation satellite
May 1	Gary Power's U-2 plane shot down over USSR
May 15	USSR launches unpiloted Vostok
May 16	Big Four meet in Paris; Eisenhower-Khrushchev summit canceled in wake of U-2 incident
May 24	U.S. launches *Midas 2* (Missile Defense Alarm System) on Atlas-Agena rocket, the first early-warning missile detection satellite
June 22	U.S. launches the first multiple payload spacecraft, *Transit/Solrad*
July 23	USSR launch of *Vostok* craft with two dogs fails to reach orbit
August 10	U.S. launches and recovers *Discoverer 13*, the first recoverable observation satellite
August 12	U.S. launches *Echo 1*, the first passive communication satellite
August 18	The CIA recovers the first satellite photographs of the Soviet Union from the*Discoverer 14* satellite.
August 19	USSR launches *Korabl* with two dogs, the first successful recovery of orbited animals
autumn	USSR and China split quietly widens at Moscow conference of 81 Communist parties
September 22 to October 13	Khrushchev to New York City for UN emergency session regarding Congo; famous shoe pounding; Eisenhower and and heads of state of Bulgaria, Cuba, Czechoslovakia, Ghana, Hungary, India, Indonesia, Poland, Romania, UAR and Yugoslavia attend
October 4	U.S. launches *Courier 1B*, the first delayed repeater communications satellite (comsat)
October 24	Soviet "Nedelin" disaster: explosion at cosmodrome
December 1	USSR launches *Vostok* with two dogs
December 22	USSR launch of *Vostok* with dog fails to reach orbit

1961

January 20	Kennedy calls on "both sides" to "explore the stars" together in his inauguration address[1]
January 30	Kennedy, in State of the Union address, suggests U.S. and USSR "help themselves as well as other nations by removing...[space] endeavors from the bitter and wasteful competition of the Cold War."[2]
January 31	U.S. launches *Samos 2*, the first successful radio-transmission reconnaissance satellite (recon sat)
February 12	USSR launches *Venera* (Venus) probe to Venus but loses contact with it February 27
February 21	USSR launches *Sputnik 5*, the first orbital platform launch
March 9	USSR launches *Vostok* with dog

March 13	Kennedy announces Alliance for Progress, a 10-point intra-American cooperation program
March 23	Soviet cosmonaut Bondarenko killed during training in ground fire
March 25	USSR launches *Vostok* with dog
April 12	USSR launches *Vostok 1*, orbiting Gagarin, the first man in space
April 15	Bay of Pigs: failed U.S.-supported invasion of Cuba
May 5	U.S. launches *Freedom 7*, putting Shepard into suborbital flight
May 25	Kennedy announces U.S. will put a man on the moon by the end of the decade
June 3-4	Kennedy and Khrushchev meet in Vienna; Kennedy reports "most somber talks" about Berlin
June 29	U.S. launches *Transit 4A*, the first satellite to use isotope power
July 21	U.S. launches *Liberty Bell 7*, putting Grissom into suborbital flight
August 6	USSR launches *Vostok 2*, orbiting Titov
August 13	Third Berlin crisis: Berlin Wall construction begun, Khrushchev demands U.S., France and Great Britain sign a peace treaty with East Germany by the end of the year
September 1	USSR resumes atmospheric nuclear tests, breaking moratorium
September 25	Kennedy addresses the UN General Assembly calling on "all nations" to cooperate in space
October 17	Sino-Soviet split becomes public at 22nd CPSU Congress
October 21	U.S. Deputy Secretary of Defense Gilpatric publicly dismisses the "missile gap," saying U.S. has a growing nuclear advantage
November 27	U.S. tests first Saturn I rocket
November 29	U.S. recovers chimpanzee "Enos" from a Mercury two-orbit flight
December 20	UN Resolution 1721 (XVI) adopted unanimously, applies international law to space exploration, expands COPUOS membership

1962

February 20	U.S. launches *Friendship 7*, orbiting Glenn three times
February 21	Khrushchev writes Kennedy, suggesting "our two countries unite their [space] efforts"
March 7	U.S. launches *OSO-1*, the first orbiting solar observatory Kennedy proposes specific cooperative steps in letter to Khrushchev
March	New COPUOS meets for the first time
March 16	USSR launches *Cosmos* (Space) 1 (designation for military satellites) from Kapustin Yar
April 25	U.S. resumes atmospheric nuclear tests

April 26	U.S. launches *Ariel 1* in cooperation with the Great Britain for ionosphere investigations; the first internationally cooperative satellite
	USSR launches the recoverable *Cosmos 4*, its first military espionage satellite; *Kosmos 5* (a non-espionage, military satellite) follows on May 28 (by December 23, 1969 313 more *Cosmos* satellites are orbited)[3]
May 24	U.S. launches *Aurora 7*, orbiting Scott Carpenter three times
June 7	USSR submits proposal to COPUOS making reconnaissance satellites illegal
June 8	USSR Academy of Sciences and NASA sign agreement on limited sharing of data
July 10	U.S. launches *Telstar 1*, AT&T's first satellite; first live transatlantic TV transmissions
July 28	USSR launches *Cosmos 7*, its second military espionage satellite
August 11	USSR launches *Vostok 3*, orbiting Nikolaev
August 12	USSR launches *Vostok 4*, only 23 hours and 32 minutes after launching *Vostok 3*
August 24	U.S. and USSR announce failure to reach a nuclear test ban after 39 months of negotiations
August 27	U.S. launches *Mariner 2*; first satellite to return close-up data from Venus
August 31	U.S. passes Communications Satellite Act (Comsat Act)
September 27	USSR launches *Cosmos 9* recon sat
October 3	U.S. launches *Sigma 7*, a Mercury spacecraft, orbiting Schirra six times
October 16-28	Cuban Missile Crisis; October 22 Kennedy announces blockade
October 17	USSR launches *Cosmos 10* recon sat
October 31	U.S. launches *Anna 1B*, the first geodetic satellite
November	Khrushchev allows the publication of Solzhenitsyn's *A Day in the Life of Ivan Denisovich*
November 1	USSR launches *Mars 1* for a flyby of Mars but loses contact with it March 21, 1963
December 14	UN Resolution 1802 (XVII) adopted unanimously
December 22	USSR launches *Cosmos 12* recon sat

1963

March 21	USSR launches *Cosmos 13* recon sat
April 16	USSR submits a second proposal to COPUOS making recon sats illegal
May 9	U.S. launches *Project West Ford*, the controversial copper dipole experiment
May 15	U.S. launches *Faith 7*, orbiting Gordon Cooper 22 times (last Mercury flight)
June 14	USSR launches *Vostok 5* with Valery Bykovsky, who sets longevity record of five days.

June 16	USSR launches *Vostok 6*, with Valentina Tereshkova, the first woman in space
	Vostok 6 and *Vostok 5* pass within 5 km of each other
June 20	U.S. and USSR conclude "hotline" agreement, establishing a direct telegraphic link
July 26	U.S. launches *Syncom 2*, the first operational geostationary comsat
summer	Sino-Soviet conflict flares
August 5	Treaty Banning Nuclear Weapon Tests in the Atmosphere, Outer Space and Under Water (also known as the Limited or Partial Test Ban) signed
September 20	Kennedy proposes a joint U.S.-Soviet piloted moon mission in address to the UN GA
October 9	Kennedy approves selling 150 million bushels of grain to USSR
October 17	UN Resolution 1884 (XVIII) passes unanimously, banning nuclear weapons from space
	U.S. launches *Vela Hotel*, the first nuclear explosion detection satellite
November 22	Kennedy assassinated
November 27	U.S. launches *Centaur 2*, the first hydrogen rocket to orbit the Earth
December 13	UN Resolutions 1962 and 1963 (XVIII) adopted unanimously, establishes beginnings of space law
1964	U.S. and 18 other nations form Intelsat, a satellite communications consortium; by April 1969 it has 68 nations as members
March 21	Chinese Communist Party issues public manifesto calling for all communists to repudiate the Soviets' leadership of the world socialist movement
June 1	U.S. and USSR sign a consular treaty, their first bilateral treaty since 1917; its implementation is not begun until January 13, 1968 (after heated debate)
June 4	U.S. *Transit 5C* completes the first operational satellite navigation system
July 28	U.S. *Ranger 7*, which takes the first close-up pictures of the moon
August 10	Gulf of Tonkin Resolution passes
October 12	USSR launches *Voskhod* (Rising) 1, first three-man flight, first doctor in space
October 14	Presidium of the Central Committee CPSU deposes Khrushchev
October 16	China explodes its first Atomic bomb
October 31	U.S. astronaut Freeman is killed during training; he is the first astronaut killed on duty
November 28	U.S. launches *Mariner 4*, which makes the first fly-by of Mars in July 1965

1965

Fiscal Year	For the first time in its history NASA's budget is less than its previous year's
February	U.S. begins bombing North Vietnam while Kosygin visits North Vietnam
February 16	U.S. launches *Pegasus 1*, the first micrometeorite satellite
February 20	U.S. launches *Ranger 8* which impacts on moon; *Ranger 9* does the same March 21
March 8	U.S. Marine detachment sent to Da Nang; the first organized U.S. combat unit to Vietnam
March 18	USSR launches *Voskhod 2*, Leonov completes the first space walk, as televised in USSR
March 23	U.S. launches *Gemini 3*, the first maneuverable spacecraft, with Grissom and Young
April 5	U.S. launches *Snapshot 1*, which includes the first nuclear reactor in space
April 6	U.S. launches *Early Bird*, the first geosynchronous commercial comsat (type "Intelsat I") over the Atlantic
April 23	USSR launches *Molniya* (Lightning) *1*, probably its first communications satellite
April 28	U.S. intervenes in Dominican Republic to prevent a communist takeover
June 3	U.S. launches *Gemini 4*, White makes the first American spacewalk
July 16	USSR's new *Proton* rocket lifts a record 26,000 pounds into orbit, launching *Proton 1*, which conducts the first satellite cosmic ray measurements
August 6	U.S. passes the Voting Rights Act of 1965, to help and encourage blacks to vote
August 21	U.S. launches *Gemini 5*, orbiting Cooper and Conrad
November 16	USSR launches *Venera 3*, the first satellite to impact on Venus
December 4	U.S. launches *Gemini 7*, sets record with 206 piloted orbits; U.S. has 1353 man-hours in space, USSR has 507[4]
December 15	U.S. launches *Gemini 6*, flies within one foot of *Gemini 7* for eight hours
December 21	UN Resolution 2130 (XVIII) adopted, calling for international space law

1966

	USSR founds Interkosmos to coordinate cooperative space ventures with socialist nations
	France leaves NATO; De Gaulle visits Moscow in June
January 1	UN Charter amended, expanding the Security Council from 11 to 15
January 14	Korolev dies during botched surgery
January 31	USSR *Luna 9* survives landing on the moon--the first successful soft lunar landing

March 16	U.S. launches *Gemini 8*, which conducts the first piloted-unpiloted craft linkup; mission aborted early
March 31	USSR launches *Luna 10*, first satellite to orbit the moon
April 8	U.S. launches its first Orbiting Astrophysical Observatory (*OAO*)
May 30	American *Surveyor 1* soft lands on the moon at 8 miles an hour
June 3	U.S. launches *Gemini 9*, conducts two rendezvous's with Agena target vehicles; first ship-to-ship spacewalk
June 25	USSR launches *Cosmos 122*, probably its first weather satellite
July 18	U.S. launches *Gemini 10*, orbiting Young and Collins, who conduct the first docked spacecraft maneuver with the unmanned *Agena 8*
August 10	U.S. *Orbiter 1*, first satellite to orbit the moon and return lunar photographs
September 12	U.S. launches *Gemini 11*, orbiting Conrad and Gordon, rendezvous with Agena
October 26	U.S. launches *Lani Bird*, first Intelsat II comsat, over the Pacific
November 11	U.S. launches *Gemini 12*, the last Gemini flight
December 6	U.S. launches *ATS-1*, the first Applications Technology Satellite Johnson and Kosygin meet in Glassboro, New Jersey
December 19	UN Resolution 2222 (XXI) adopted; includes a draft space law treaty
December 21	USSR launches *Luna 13*, conducts the first lunar surface bearing test
1967	
January 11	U.S. launches *Pacific Two* comsat over the Pacific
January 25	USSR launches *Cosmos 139*, the first test of its Fractional Orbital Bombardment System to launch ICBMs over South Pole to avoid the Ballistic Missile Early Warning System
January 27	U.S.'s *Apollo 204* catches fire, killing Chaffee, Grissom and White in launch rehearsal
	UN approves Treaty on the Use Of Outer Space
March 10	USSR launches unpiloted *Zond* spacecraft which fails in circumlunar flight; three more failures follow on April 8 & November 21, 1967 and April 22, 1968
March 22	U.S. launches the second Atlantic geosynchronous satellite
April 17	U.S. launches *Surveyor 3*, which collects moon samples in the first lunar trenching
April 23	USSR launches *Soyuz 1*, its first piloted mission since March 1965; the capsule crashes, killing cosmonaut Komarov during apparent mission abort; first reported cosmonaut death
April 25	U.S., USSR, Great Britain and 57 other nations sign the Outer Space Treaty; 80 more sign in ensuing weeks (France notably does not sign)
June 5-10	Arab-Israeli Six-Day War; Israel takes Sinai Peninsula, W. Bank, Gaza Strip, Golan Heights

June 8	U.S. uses superpower hotline for the first time, notifying USSR of use of aircraft near Israel
June 12	USSR launches *Venera 4*, which is the first satellite to probe Venus' atmosphere
June 14	U.S. launches *Mariner 5* which flies by Venus on October 19
June 23-25	Johnson and Brezhnev meet at Glassboro, NJ; the fourth meeting of the superpowers' heads of state since the end of World War II
July 1	U.S. launches *DODGE*, which takes the first full color picture of the Earth
August 1	U.S. launches *Lunar Orbiter 5*, which completes a map of the entire lunar surface, as begun by *Lunar Orbiter 1, 2, 3 & 4* (launched from August 10, 1966 to May 4, 1967)
September 8	U.S. launches *Surveyor 5*, which does the first advanced chemical analysis of lunar soil; *Surveyor 6 & 7* follow on November 7, 1967 and January 7, 1968
September 27	U.S. launches the third Pacific geosynchronous satellite
October 27-29	USSR launches *Cosmos 186 & 188*, which conduct the first automatic space docking
November 4	U.S. and USSR agree to establish direct domestic flights; flights begin July 1958
November 9	U.S. successfully tests Saturn 5, orbiting a record 278,699 pounds and launching *Apollo 4*, which conducts the first lunar-velocity reentry
December 19	UN Resolution 2345 (XXI) adopted; addresses rescue and return of astronauts and space objects

1968

January 23	North Korea seizes U.S. Navy's *Pueblo*; its 82 crewmembers are returned December 22
January 30	North Vietnamese Tet Offensive inflicts damaging blows to U.S., South Vietnamese forces
March 2	USSR launches *Zond 4* (a modified Soyuz spacecraft), the first of four successful unpiloted circumlunar probes launched between 1968 and 1970
April 22	U.S. and USSR sign Agreement on the Rescue and Return of Astronauts and Space Objects
April 23	USSR launches *Soyuz 1*, the first flight of the new spacecraft
April 4	Martin Luther King, Jr. is assassinated; he receives the Nobel Peace Prize December 10
	U.S. launches *Apollo 6*, the second Saturn 5 test
June 5	Senator Robert F. Kennedy (D - NY) assassinated
July 1	U.S., USSR and 60 other nations sign the nuclear Non-Proliferation Treaty
June 28	USSR launches *Zond 5*, the first successful recovery of circumlunar animals; *Zond 6* repeats the mission November 10
October 11	U.S. launches *Apollo 7*, the first piloted Apollo flight, with

	three astronauts
October 20	USSR tests the first co-orbital ASAT device
October 25	USSR launches *Soyuz 2*, unpiloted
October 26	USSR launches *Soyuz 3*, its first piloted flight since the Komarov accident
December 7	U.S. launches *OAO-2*, the first point-stabilized satellite
December 18	U.S. launches its first Intelsat III-type comsat over the Atlantic; another is launched over the Pacific in February
December 21	U.S. launches *Apollo 8*, the first piloted circumlunar flight; orbits the moon ten times and returns the first live lunar television broadcast
1969	Conference on Security and Cooperation in Europe USSR and West Germany sign their first treaty
January 14	USSR launches *Soyuz 4*, one-man flight
January 15	USSR launches *Soyuz 5*, three-man flight; docks with *Soyuz 4* for four hours, transferring two cosmonauts (the first docking between piloted spacecraft and first crew transfer)
March 2– September	USSR and China battle each other in open troop combat along Sino-Soviet border
March 3	U.S. launches *Apollo 9*, testing the entire Apollo configuration (spacecraft, command and service module and lunar module) with three astronauts in Earth orbit
March 27	USSR launches *Meteor 1*, a weather satellite
May 18	U.S. launches *Apollo 10*; three astronauts practice separation and re-docking
June	Soviet rocket (N-1) intended for piloted lunar mission destroyed in fueling accident
July 16	U.S. launches *Apollo 11*; Neil Armstrong sets foot on the moon July 20 in the first piloted lunar landing and return
July 31 & Aug. 5	U.S. *Mariner 6* and *7* probes fly past Mars
Oct. 11, 12 & 13	USSR launches *Soyuz 6, 7* and *8*, the first time there were three piloted spacecraft aloft simultaneously
November 3	Nixon announces Nixon Doctrine, stressing America's nuclear shield and allies' need to supply manpower for defense
November 14	U.S. launches *Apollo 12*, 2nd successful piloted moon landing
1970	USSR begins developing new Anti-Ballistic Missile defense around Moscow[5]
April 11	U.S. launches *Apollo 13*; three astronauts fail to land on moon, return safely
June 1	USSR launches *Soyuz 9*; its two cosmonauts orbit for 18 days
July 16	USSR launches *Venera 7*, which conducts the first soft landing on Venus
August 7	Egypt and Israel agree to a cease-fire after more than a year's fighting in the Suez Canal
September 12	USSR launches *Luna 16*, returns first lunar sample

November 10   USSR launches *Luna 17*, first automated lunar roving lab

1971   U.S. government admits funding Radio Free Europe and Radio
                   Free Liberty for the first time (founded 1949 & 1951 and
                   broadcast to East Europe and the USSR, respectively)
January 21   NASA and USSR Academy of Sciences agree to exchange
                   moon samples
January 31   U.S. launches *Apollo 14*, third successful piloted moon landing
April 19   USSR launches *Salyut 1*, first space station; stays aloft until
                   October 11, 1971
April 22   USSR launches *Soyuz 10*; cosmonauts fail to board *Salyut 1*
June 6   USSR launches *Soyuz 11*; three cosmonauts board Salyut 1, stay
                   for 24 days, then suffocate on reentry (June 29); they wore no
                   spacesuits; first astronauts to die in space
June 19   USSR launches *Mars 2*, the first craft to impact Mars
June 28   USSR launches *Mars 3*, the first craft to soft land on Mars
July 15   Nixon announces he will visit Peking
July 26   U.S. launches *Apollo 15*; fourth successful moon landing;
                   astronauts ride on the first piloted moon rover
October 25   U.S., USSR, Great Britain and other nations sign the Seabed
                   Treaty, banning nuclear weapons from the ocean floor
November 13   *Mariner 9* enters orbit around Mars, its first artificial satellite;
                   maps Mars

1972
January 5   Nixon approves the development of the space shuttle
February 17-28   Nixon visits Peking
March 3   U.S. launches *Pioneer 10*, the first probe to return close-up
                   photos of Jupiter, fly by Mercury and Venus and escape the
                   solar system
March 27   USSR launches *Venera 8*, which conducts the first Venus soil
                   analysis
April 16   U.S. launches *Apollo 16*; fifth successful moon landing
May 22-30   Nixon becomes the first U.S. President to visit Moscow in
                   summit with Brezhnev
May 24   U.S. and USSR sign Agreement Concerning Cooperation in the
                   Exploration and Use of Outer Space for Peaceful Purposes,
                   preparing for the Apollo-Soyuz joint mission
May 25   U.S. and USSR sign Agreement on the Prevention of Incidents
                   on and over the High Seas
May 26   Carter and Brezhnev sign Treaty on the Limitation of Anti-
                   Ballistic Missile Systems and SALT I (Strategic Arms
                   Limitations Talks)
June 17   Watergate break-in
July 5   USSR purchases a record 8.5 million tons of U.S. grain
July 23   U.S. launches *Landsat 1* (the first ERTS, or Earth Resources
                   Technology Satellite) to map Earth

July 29	USSR attempt to launch *Salyut 2-1* fails, destroying the station
August 21	U.S. launches *Orbiting Astrophysical Observatory 3*, an x-ray observatory
December 7	U.S. launches *Apollo 17*; sixth and final moon landing; record 22 hours on surface

1973
January 27	U.S., South Vietnam, North Vietnam and Viet Cong sign formal peace agreement
April 3	USSR launches unpiloted *Salyut 2*; it is never boarded
April 6	U.S. launches *Pioneer 11*, which returns the first close-up color pictures of Saturn
May 11	USSR launches *Cosmos 557*, apparently a military space station; it was never boarded
May 14	U.S. launches *Skylab 1*; three piloted spacecraft dock with it
May 25	U.S. launches three astronauts in *Skylab 2* to *Skylab*; they conduct critical spacewalk repairs and return on June 22
June 10	U.S. launches *Explorer 49*, the first radio telescope satellite
June 30	U.S. ends mandatory draft for the first indefinite period since 1948
July 28	U.S. launches *Skylab 3* to *Skylab*
September 27	USSR launches *Soyuz 12*, its first successful piloted flight since June 1970
October 18 to March 18, 1974	OPEC oil embargo
November 7	Congress overrides Nixon's veto, passes the War Powers Act
November 10	U.S. launches *Mariner 10*, which returns the first close-up pictures of Venus and Mercury
November 11	Israel and Egypt agree to cease-fire, ending Yom Kippur War (it started on October 6, 1973)
November 16	U.S. launches *Skylab 4* crew of three to *Skylab*
December 18	USSR launches *Soyuz 13*
December 29	USSR launches *Cosmos 626*,[6] its first radar ocean surveillance

1974
	U.S. and USSR interests clash in Angola
January 10	Secretary of Defense Schlesinger announces "limited strategic strike options" policy
May 17	U.S. launches *SMS-1*, the first geostationary weather satellite
May 18	India explodes its first atomic bomb
May 30	U.S. launches *ATS-6*, the most powerful comsat to date
June 24	USSR launches *Salyut 3*, a military space station; one piloted spacecraft docks with it
June 27 to July 3	Nixon visits USSR for summit with Brezhnev
July 3	USSR launches *Soyuz 14*, crew docks with *Salyut 3* until July 19; first successful return from a *Salyut* space station
August 9	Nixon resigns
August 26	USSR launches *Soyuz 15*, which fails in docking with *Salyut 3*

December 2	USSR launches *Soyuz 16*
December 10	U.S. launches *Helios 1*, the first solar probe
December 26	USSR launches *Salyut 4*; three piloted spacecraft dock with it

1975

January 3	U.S. law passed linking Soviet Most Favored Nation (MFN) trading status with Jewish emigration
January 10	USSR launches *Soyuz 17*, crew docks with *Salyut 4* until February 9
April 5	USSR's *Soyuz 18-1* fails to reach orbit; it is the world's first known aborted piloted flight
April 29	South Vietnam surrenders; the last Americans are evacuated from Saigon
May 24	USSR launches *Soyuz 18*, crew docks with *Salyut 4* until July 26 (sixty-three days in space)
June 8	USSR launches *Venera 9*, which takes the first photographs of Venus' surface and becomes Venus' first artificial satellite
July 15	USSR launches *Soyuz 19*
	U.S. launches its last *Apollo* spacecraft
July 17	Apollo-Soyuz Test Project: Apollo and Soyuz dock in space, American and Soviet astronauts board each others' spacecraft
August 1	U.S., Canada, USSR and 32 other European nations sign the Helsinki Agreement on human rights and increased exchanges
August 20	U.S. launches *Viking 1* which successfully soft lands on Mars and returns the first pictures of the Martian surface; *Viking 2* follows on September 9
October 9	Andrei Sakharov, Soviet dissident and physicist is awarded the Nobel Peace Prize; in January 1980 Soviet officials arrest him and force him into exile
October 16	U.S. launches the 1st geostationary environmental sat, *GOES-1*
November 17	USSR launches *Soyuz 20*
November 19	USSR launches *Cosmos 782*, carrying American biological experiments

1976

January 15	U.S. launches *Helios 2* solar probe
February 18	Ford restructures U.S. foreign intelligence, largely in response to mail opening by the CIA
February 19	U.S. launches *Marisat 1*, the first commercial maritime comsat
May 4	U.S. launches *Lageos*, a satellite to study plate tectonics and the Earth's shape
May 28	U.S. and USSR sign a treaty restricting the use of nuclear explosives for peaceful purposes
June 22	USSR launches *Salyut 5*, a military space station; two piloted spacecraft dock with it
July 6	USSR launches *Soyuz 21*, crew docks with *Salyut 5* until August 24
September 15	USSR launches *Soyuz 22*, a week-long photographic mission

October	U.S.-Soviet SALT II talks stall over disagreements regarding bombers and cruise missiles
October 14	USSR launches *Soyuz 23*, which fails in docking with *Salyut 5* (fourth docking failure out of nine tries in five years)
1977	U.S. begins developing the Miniature Homing Vehicle ASAT device using the F-15 aircraft; the program is cancelled in 1988
	U.S. and USSR interests clash in Ethiopia
February 7	USSR launches *Soyuz 24*, crew docks with *Salyut 5* until February 25
May 11	U.S. and USSR sign Agreement on Cooperation in the Area of Manned Space Flight
August 12	U.S. launches *HEAO-1*, its first High-Energy Astronomy Observatory satellite
	U.S. space shuttle prototype *Enterprise* conducts its first free flight test
August 20	U.S. launches *Voyager 2*
September 5	U.S. launches *Voyager 1* [sic]
September 29	USSR launches *Salyut 6*, a civilian space station; 16 piloted spacecraft dock with it
October 9	USSR launches *Soyuz 25*; crew returns after three days and failing to dock with *Salyut 6*
October 17	USSR adopts new constitution, replacing Stalin's 1936 constitution
November 22	U.S. launches its first European Space Agency (ESA) satellite, *Meteosat 1*
December 10	USSR launches *Soyuz 26*; crew sets endurance record of 96 days on *Salyut 6*; stay includes first double docking, spacecraft exchange and in-flight resupply and refueling
1978	Carter initiates U.S.-Soviet talks to ban space weapons; talks fail in 1979
January 10	USSR launches *Soyuz 27*; it docks with *Salyut 6* (where *Soyuz 26* still is), the first two-ship docking
January 20	USSR launches *Progress 1*, which is the first craft to automatically transfer propellant
January 26	U.S. launches the *International Ultraviolet Explorer*, an astrophysical observatory and cooperative project with ESA
March 2	USSR launches *Soyuz 28* crew (including Czech cosmonaut as part of "Interkosmos" program) docks with *Salyut 6*; the first iinternational crew launch
May 20	U.S. launches *Pioneer-Venus 1* probe; it begins orbiting Venus on December 4
June 2-25	U.S. and USSR cooperate in rocket measurements of atmosphere at Wallops Island, VA
June 15	USSR launches *Soyuz 29*, crew sets endurance record of 140 days on *Salyut 6*

June 27	U.S. launches *Seasat 1*, the first ocean resources satellite, to study the oceans
	USSR launches *Soyuz 30*, crew (including Pole) docks with *Salyut 6*
August 8	U.S. launches *Pioneer-Venus 2* spacecraft; its 4 probes descend through Venus' atmosphere
August 12	U.S. launches *ISEE-3*, the International Sun-Earth Explorer-3 to study solar winds; in June 1982 NASA changes the satellite's name and mission to *ICE*, the International Cometary Explorer, which makes the first rendezvous with a comet
August 26	USSR launches *Soyuz 31*, crew (including East German) docks with *Salyut 6*
October 13	U.S. launches *Tiros N*, first of a new generation of weather sats
December 15	U.S. and China establish formal diplomatic ties

1979

February 11	Ayatollah-led revolution overthrows the Shah of Iran, seizes U.S. embassy for several hours
February 25	USSR launches *Soyuz 32*, crew sets endurance record of 175 days on *Salyut 6*
March 5	U.S.'s *Voyager 1* makes its closest approach to Jupiter on its way to Saturn
March 26	Israel, Egypt and U.S. sign peace treaty, ending the formal state of war which had existed between Israel and Egypt since May 15, 1948
April 10	USSR launches *Soyuz 33*, crew (including Bulgarian) fails to dock with *Salyut 6*
June 6	USSR launches *Soyuz 34*; unpiloted flight supplies *Salyut 6*
June 18	Carter and Brezhnev sign SALT II agreements
July 9	U.S.'s *Voyager 2* makes its closest approach to Jupiter
August 31	U.S. State Dept. confirms the discovery of "unacceptable" Soviet combat brigade in Cuba
September 1	U.S.'s *Pioneer 11* becomes the first probe to reach Saturn
September 20	U.S. launches *HEAO-3* (*Einstein*) x-ray telescope
November 4	U.S. hostages and embassy seized in Tehran; 52 Americans are held for 444 days
December 12	NATO announces it will deploy new cruise and medium-range ballistic missiles in Europe
December 27	Soviets invade Afghanistan

1980

February 14	U.S. launches *Solar Max*, the Solar Maximum Mission; the satellite fails after 11 months
April 9	USSR launches *Soyuz 35*, crew sets endurance record of 185 days on *Salyut 6*
April 25	Eight U.S. servicemen killed in failed hostage rescue
May 26	USSR launches *Soyuz 36*, docks with *Salyut 6*, crew includes

	Hungarian
June 5	USSR launches *Soyuz T-2*, docks with *Salyut 6*
July 19	Moscow Olympics begin; U.S. leads boycott of almost sixty nations
July 23	USSR launches *Soyuz 37*, crew includes a Vietnamese who Soviets claimed was the first to shoot down a B-52 over North Vietnam (USAF disputed this claim), docks with *Salyut 6*
September 18	USSR launches *Soyuz 38*, crew includes the first black in space, a Cuban
November 27	USSR launches *Soyuz T-3*, docks with *Salyut 6*
December 6	U.S. launches the first fifth generation (Intelsat V) comsat

1981

March 12	USSR launches *Soyuz T-4*, docks with *Salyut 6*
March 22	USSR launches *Soyuz 39*, docks with *Salyut 6*
March 30	Attempted assassination of Reagan
April 12	U.S. launches STS-1, space shuttle *Columbia*, the first shuttle flight and the first reusable spacecraft
May 14	USSR launches *Soyuz 40*, the last craft to dock with *Salyut 6*
November 12	U.S. launches STS-2, space shuttle *Columbia*, the first repeat use of a spacecraft

1982

March 22	U.S. launches STS-3, space shuttle *Columbia* for its 3rd mission
April 19	USSR launches *Salyut 7*; 10 piloted spacecraft dock with it
May 13	USSR launches *Soyuz T-5*, docks with *Salyut 7*; crew sets duration record of 211 days
June 24	USSR launches *Soyuz T-6*, docks with *Salyut 7*; crew includes the first French astronaut
June 27	U.S. launches STS-4, space shuttle *Columbia* for its final test mission; launches the first shuttle DoD payload
July 16	U.S. launches *Landsat 4*, a new generation of Earth-resources satellite
August 19	USSR launches *Soyuz T-7*, docks with *Salyut 7*
September 29	1200 U.S. Marines are sent to Lebanon as a part of a multi-national peacekeeping force
October 8	Polish government outlaws Solidarity; Reagan announces repeal of Poland's MFN trading status the next day
November 10	Brezhnev dies, Andropov (KGB head for the last 15 years) becomes General Secretary
November 11	U.S. launches STS-5, space shuttle *Columbia* on its first full mission; carries a record 4-person crew and launches 2 comsats

1983

January 25	U.S. launches *International Infrared Astronomy Satellite* (*IRAS*)
March 14	Reagan calls the USSR the "evil empire"

March 23	Reagan, in nationally televised "Star Wars" address, calls for ballistic missile defense system, initiating the Strategic Defense Initiative (SDI)
April 4	U.S. launches STS-6, space shuttle *Challenger*, on its first flight; crew deploy the first *Tracking and Data Relay Satellite* (*TDRS*)
April 20	USSR launches *Soyuz T-8*, which fails to dock with *Salyut 7*
June 18	U.S. launches STS-7, space shuttle *Challenger* with a record five-person crew, including America's first female astronaut, Sally Ride
June 28	USSR launches *Soyuz T-9*, docks with *Salyut 7* and conducts the first construction in space
August 30	U.S. launches STS-8, space shuttle *Challenger*, the first night-launch of a space shuttle; crew includes the first black American astronaut
September 1	USSR shoots down Korean Airlines flight #007, killing 269
September 26	USSR attempts to launch *Soyuz T-10A*, but crew escapes from capsule fire before launch
October 23	Terrorist bombing kills 241 U.S. soldiers in Lebanon
October 25	U.S. invades Grenada
November 28	U.S. launches Space Shuttle *Columbia*, including a German, the first non-American astronaut on a U.S. mission; carries the first Spacelab biological experiment module

1984

January 25	Reagan calls on NASA to build a permanently manned space station within ten years, inaugurating Space Station Freedom
February 3	U.S. launches STS 41-B, Space Shuttle *Challenger*; crew tests the Manned Maneuvering Unit (MMU), the first such device for spacewalks; first shuttle landing at Kennedy Space Center in Florida (usual landing site is Edwards Air Force Base, California)
February 8	USSR launches *Soyuz T-10*, docks with *Salyut 7*; crew sets duration record of 237 days
February 9	Andropov dies, Chernenko (Brezhnev's chief aid for 30 years) becomes General Secretary
March 1	U.S. launches *Landsat 5*, the fifth U.S. Earth-resources satellite
April 3	USSR launches *Soyuz T-11*; crew includes the first Indian in space
April 6	U.S. launches STS 41-C, Space Shuttle *Challenger*, which conducts the first in-space satellite repair, fixing *Solar Max*
July 17	USSR launches *Soyuz T-12*, which docks with *Salyut 7*
August 30	U.S. launches STS 41-D, Space Shuttle *Discovery*, the third Shuttle's 1st and the overall 100th manned space flight; crew includes first non-professional astronaut; deploys a record three satellites
September	Reagan meets with Soviet Foreign Minister Andrei Gromyko
October 5	U.S. launches STS 41-G, Space Shuttle *Challenger*, with a record

	seven-person crew
October 11	Congress approves $254.1 million for new F-15 ASAT R&D
November 8	U.S. launches STS 41-G, Space Shuttle *Discovery*, retrieves two disabled satellites
1985	USSR launches *Cosmos 1667*, carrying U.S.-built biomedical equipment
January 24	U.S. launches STS 51-C, Space Shuttle *Discovery* on a dedicated DoD mission
March 12	Chernenko dies, Gorbachev becomes General Secretary; Gorbachev's policies evolve from *uskorenie* (acceleration, 1985), to *perestroika* (rebuilding) and *glasnost* (openness, starting 1986) to *demokratizatsiya* (democratization, starting December 1988) U.S. and USSR resume arms control talks
April 12	U.S. launches STS 51-D, Space Shuttle *Discovery*; crew includes U.S. Senator Garn
April 29	U.S. launches STS 51-B, Space Shuttle *Challenger*, carrying Spacelab 3
June 5	USSR launches *Soyuz T-13*, which repairs *Salyut 7*
June 17	U.S. launches STS 51-G, Space Shuttle *Discovery* including a Saudi Arabian and French astronaut; deploys three comsats
summer	U.S.-Soviet relations marred by reports of spying at American embassies
July 29	U.S. launches STS 51-F, Space Shuttle *Challenger*, carrying Spacelab 2
August 27	U.S. launches STS 51-I, Space Shuttle *Discovery*, repairs *Leasat 3*
September 17	USSR launches *Soyuz T-14*, docks with *Salyut 7*
October 3	U.S. launches STS 51-J, *Atlantis*-the fourth Space Shuttle, on a DoD mission
October 30	U.S. launches STS 61-A, Space Shuttle *Challenger* with a record eight-person crew
November 19-21	Reagan and Gorbachev meet in Geneva in what Reagan calls a "fresh start"
November 21	U.S. and USSR sign agreement on "Contacts, Exchanges, and Cooperation in Scientific, Technical, Educational, Cultural. and other fields"
November 27	U.S. launches STS 51-B, Space Shuttle *Atlantis*, includes the first Mexican astronaut
1986	
January 12	U.S. launches STS 61-C, Space Shuttle *Columbia*, after several delays; crew includes U.S. Representative Nelson
January 24	U.S.'s *Voyager 2* returns the first close-up photographs of Uranus, reaches Neptune in 1989
January 28	U.S. STS-51L, Space Shuttle *Challenger*, explodes, killing seven astronauts
February 20	USSR launches space station *Mir*'s primary module

March 13	USSR launches *Soyuz T-15,* which acts as the first inter-station ferry, docking with *Salyut 7* and *Mir*
April 14	U.S. bombs Libya in retaliation for supporting terrorism
April 18	U.S. Titan 34D rocket, carrying a recon sat, explodes
April 26	Severe nuclear accident at Chernobyl nuclear reactor in Ukraine
May 3	U.S. suffers its third consecutive launch failure: a Delta 3914 rocket fails to achieve orbit
May 27	Reagan announces the U.S. will no longer comply with SALT II
September 5	U.S. launches an SDI payload on a Delta rocket
October 11-12	Reagan and Gorbachev meet in Reykjavik; proposed sweeping arms reductions fail due to disagreement over SDI

1987	U.S. begins developing ground-based laser ASAT devices as a part SDI
February 5	USSR launches *Soyuz TM-2,* docks with *Salyut 7;* crew sets duration record of 326 days
March 26	U.S. Atlas-Centaur rocket fails to reach orbit
March 31	USSR launches *Kvant-1* expansion module to *Mir*
April 15	U.S. and USSR sign agreement on space cooperation
May 17	U.S. frigate *Stark,* escorting a Kuwaiti tanker, is hit by Iraqi jet, killing 37
	USSR launches *Cosmos 1887,* payload includes U.S.-built biomedical equipment
July 22	USSR launches *Soyuz TM-3,* docks with *Mir;* crew includes first Syrian in space
September to November	Yeltsin, First Party Secretary of Moscow, challenges Gorbachev's reforms as too slow; Gorbachev removes him but allows him to remain on the Central Committee
October 19	Dow Jones Industrial Average loses 22.6%, despite near-record 59 consecutive months of economic growth; worldwide investor panic ensues
December 8	Reagan and Gorbachev sign treaty banning intermediate-range nuclear forces (INF) at Washington summit; summit continues through December 10
December 21	USSR launches *Soyuz TM-4,* docks with *Mir*

1988	
February	Widespread ethnic riots in Nagorno-Karabakh, Azerbaijan
March 25	U.S. launches *San Marco D/L,* international satellite to study Earth's lower atmosphere
May 15	USSR begins withdrawing from Afghanistan
May 29 to June 2	Reagan and Gorbachev meet in Moscow--their fourth summit
June 7	USSR launches *Soyuz TM-5,* docks with *Mir;* crew includes first Bulgarian in space
July 3	U.S. cruiser *Vincennes* shoots down Iranian airliner, killing 290

August 29	USSR launches *Soyuz TM-6*, docks with *Mir*; crew includes first Afghanistani in space
September 29	U.S. launches STS-26, Space Shuttle *Discovery*; first shuttle mission after *Challenger* explosion, after several long delays
October 27	Reagan announces U.S. must destroy its newly-built, heavily bugged Moscow embassy
November	Cuba agrees to a timetable to withdraw its troops from Angola
November 4	U.S. joins the 1948 Genocide Treaty
November 26	USSR launches *Soyuz TM-6*, docks with *Mir*
December 2	U.S. launches STS-27, space shuttle *Atlantis*, on a dedicated DoD mission
December 7	Massive earthquake in Armenia
December 7-8	Gorbachev calls for disarmament at UN and initiates unilateral arms cuts in Europe

1989

	Revolutions in Bulgaria, Czechoslovakia, East Germany, Romania, Hungary and Poland
February 14	U.S. launches *GPS-1*, beginning its extensive Global Positioning System
March to May	USSR holds multi-candidate (not multi-party) elections to Congress of People's Deputies
March 13	U.S. launches STS-29, space shuttle *Discovery*
May 4	U.S. launches STS-30, space shuttle *Atlantis*; launches Venus orbiter *Magellan*
June 4	Poland holds its first free parliamentary elections in more than 44 years; Solidarity wins 92 of 100 senate seats contested
July	Reform-minded deputies (including Yeltsin and Sakharov) form inter-regional caucus within the Congress of People's Deputies; first proto-party in USSR
August 8	August launches STS-28, space shuttle *Columbia*, on a dedicated DoD mission
September 5	USSR launches *Soyuz TM-8*, docks with *Mir*
October 18	U.S. launches STS-34, space shuttle *Atlantis*, launches Jupiter probe and orbiter *Galileo*
November 9	Berlin Wall opened
November 18	U.S. launches *Cosmic Background Explorer*
November 23	U.S. launches STS-33, space shuttle *Discovery*, on a dedicated DoD mission
November 26	USSR launches *Kvant-2* expansion module to *Mir*
December 2-3	Bush and Gorbachev meet at Yalta summit
December 3	East German Communist Party leadership resigns
December 7	13 days after Czech Politburo resigns, Czech Prime Minister resigns; interim government formed includes the Civic Forum
December 19	Bulgaria announces it will hold free elections

December 19-22   Thousands killed in anti-Ceausescu riots in Romania;
                 Ceausescu is executed on the 25th and the interim government
                 announces elections for April 1990

1990
January 9   U.S. launches STS-32, space shuttle *Columbia*; crew retrieves
                 Long-Duration Exposure Facility
February 11   USSR launches *Soyuz TM-9*, docks with *Mir*
February 28   U.S. launches STS-36, space shuttle *Atlantis*, on a dedicated
                 DoD mission
March 4   Elections to Russian parliament; Democratic Russia network
                 wins 35-40% of the seats; Yeltsin is elected with 85% of the
                 vote of his Sverdlovsk district
March 11   Lithuanian parliament unanimously votes to secede from
                 USSR
March 15   Gorbachev is elected the first Soviet President by the Congress
                 of Peoples' Deputies
March 30   Estonian parliament pledges to seek independence from USSR
April 24   U.S. launches STS-31, space shuttle *Discovery*; launches Hubble
                 Space Telescope, which is subsequently found to have a
                 flawed primary mirror
May 4   Latvian parliament announces a plan for independence
May 31   USSR launches *Kristall* expansion module to *Mir*
May 31-June 3   Bush and Gorbachev agree to destroy their nations' chemical
                 weapons at Washington summit; by June 1992 their chemical
                 weapons stockpiles are reduced by 80%
June 1   U.S. and USSR agree to grant each other unconditional MFN
                 trading status
August 1   USSR launches *Soyuz TM-10*. docks with *Mir*
September   Gorbachev acquires the right to rule by decree
October 3   West and East Germany unified at midnight
October 6   U.S. launches STS-41, space shuttle *Discovery*, launches *Ulysses*
                 solar and interstellar probe
November 15   U.S. launches STS-38, space shuttle *Atlantis* on a dedicated
                 DoD mission
November 19   U.S., USSR and the 20 European members of NATO and
                 Warsaw Pact sign the Conventional Forces in Europe
                 agreement
December 2   U.S. launches STS-35, space shuttle *Columbia*
                 USSR launches *Soyuz TM-11*, docks with *Mir*
December 11   Bush signs two long-delayed treaties with the USSR, the
                 Threshold Ban Treaty and the Peaceful Nuclear Explosions
                 Treaty (originally concluded July 3, 1974/May 28, 1976)

1991
January 13   USSR intervenes in Lithuania, soon steps up military
                 maneuvers in Latvia and Estonia

January 16	U.S. begins bombing Iraq: start of the 6-week Persian Gulf War
January 20	Soviet troops storm the Latvian Interior Ministry
April 5	U.S. launches STS-37, space shuttle *Atlantis*, launches *Gamma Ray Observatory*
April 28	U.S. launches STS-39, space shuttle *Discovery* on a dedicated DoD mission
May 18	USSR launches *Soyuz TM-12*, docks with *Mir*; cosmonaut Krikalev "stranded" in space for 313 days when USSR dissolves while he is on board *Mir*
June 5	U.S. launches STS-40, space shuttle *Columbia*; carries the first Spacelab Life Sciences (SLS) mission
June 13	Yeltsin elected President of Russia with 57.4% of the vote
summer	CPSU membership has dropped by 5 million in the last 18 months, to 15 million
July 18	Gorbachev asks leaders of the seven leading industrial countries (G-7) for economic aid
July 29	Russian Federation acknowledges Lithuania as sovereign
July 31	Bush and Gorbachev sign Strategic Arms Reduction Treaty (START) at Moscow summit, the first treaty to cut long-range nuclear weapons
August 2	August launches STS-43, shuttle *Atlantis*, launches *TDRS-5*
August 15	USSR launches *Meteor 3*, carrying NASA's Total Ozone Mapping Spectrometer
August 19-21	Conservative coup against Gorbachev government fails
August 20	Planned signing of the Union Treaty is cancelled in the wake of the coup Estonia declares full independence; Latvia does the same on August 21
September 2	U.S. acknowledges Baltic states' independence
September 6	USSR acknowledges Baltic states' independence; Georgia declares independence
September 12	U.S. launches STS-48, space shuttle *Discovery*, launches *Upper Atmosphere Research Satellite*
September 27	Bush announces large unilateral arms reductions, including the removal of short-range nuclear weapons from ships and European and Asian bases; constant bomber flights end
October 2	USSR launches *Soyuz TM-13*, docks with *Mir*
October 5	Gorbachev announces USSR will go below START levels and eliminate all tactical nuclear weapons
November 6	Yeltsin issues a decree abolishing CPSU on Russian territory
November 24	U.S. launches STS-44, space shuttle *Atlantis*, launches *Defense Support Program* satellite
December 8	Russia, Ukraine and Byelorussia declare Soviet Union defunct, form the Commonwealth of Independent States (CIS)
December 21	Remaining eight former Soviet republics (excepting Georgia and the Baltic republics) join the CIS
December 25	Gorbachev resigns as President of the USSR; the Russian flag flies over the Kremlin

1992

January 2	Russian First Deputy Prime Minister and Economics Minister Gaidar initiates major retail price liberalization program; wholesale prices liberalized April 1, 1992
January 22	U.S. launches STS-42, space shuttle *Discovery*, carrying International Microgravity Lab-1
March 17	Russia launches *Soyuz TM-14*, docks with *Mir*; first piloted CIS mission
March 24	U.S., Russia, Byelorussia, Ukraine, Georgia, 15 other NATO countries and 5 Warsaw Pact countries sign an Open Skies Agreement allowing approved overflights
	U.S. launches STS-45, space shuttle *Atlantis*
April 1	Bush announces G-7 $24 billion aid package to former USSR
May 7	U.S. launches STS-49, space shuttle *Endeavour*, first flight of shuttle replacing *Challenger*
June 17	Bush and Yeltsin sign agreements broadening space cooperation at Washington summit
	Yeltsin becomes the first Russian leader to address a Joint Session of Congress
June 25	U.S. launches STS-50, space shuttle *Columbia* for 14-day mission, the shuttle's longest
July 27	Russia launches *Soyuz TM-15*, docks with *Mir*
July 31	U.S. launches STS-46, space shuttle *Atlantis*
September 12	U.S. launches STS-47, space shuttle *Endeavour*; crew includes first black female astronaut
September 18	Pentagon confirms the existence of the National Reconnaissance Office (est. Aug. 25, 1960)
September 25	U.S. launches *Mars Observer* probe
October 22	U.S. launches STS-52, space shuttle *Columbia*
December 2	U.S. launches STS-53, space shuttle *Discovery*

1993

	North Korea publicly considers renouncing the Non-Proliferation Treaty, beginning dispute with UN and U.S.
January 3	Bush and Yeltsin sign START 2, proposing to cut U.S. warheads to 3500, (former) Soviet to 3000
January 13	U.S. launches STS-54, space shuttle *Endeavour*
January 24	Russia launches *Soyuz TM-16*, docks with *Mir*
February 21	Russia launches *Progress M-16* to resupply *Mir*; resupply missions *Progress M-17* and *Progress M-18* follow on March 31 and May 22, 1993
April 3-4	Clinton and Yeltsin meet at Vancouver summit
April 8	U.S. launches STS-56, space shuttle *Discovery*
April 15	G-7 announces a $43.4 billion dollar aid package to Russia
April 25	Russia holds referendum; 65% of voters participate, 64% of these approve early parliamentary elections
April 26	U.S. launches STS-55, space shuttle *Columbia*; carries the second German Spacelab

June 21	U.S. launches STS-57, space shuttle *Endeavour*, carrying Spacehab scientific module
July 1	Russia launches *Soyuz TM-17*; three-person crew includes French astronaut
July 10	Clinton and Yeltsin meet during G-7 Tokyo conference
August 10	Russia launches *Progress M-19* to resupply *Mir*; *Progress M-20* follows on October 11
August 21	U.S. loses contact with *Mars Observer* probe as it approaches Mars
September 2	U.S. and Russia sign agreement on international space station and Missile Technology Control Regime at first Gore-Chernomyrdin meeting
September 12	U.S. launches STS-51, space shuttle *Discovery*; deploys the first Advanced Communications Technology Satellite (ACTS)
October 18	U.S. launches STS-58, space shuttle *Columbia*; carries SLS-2; third Spacelab flight of 1993
October 26	Russia launches *Cosmos 2265*, a military calibration satellite, on a Cosmos booster
October 30	Haitian military regime prevents the U.S.- & UN-sponsored return of President Aristride
November 2	Russia launches *Cosmos 2266*, a navigation satellite (navsat) on a Cosmos booster
November 5	Russia launches *Cosmos 2267*, a photo recon sat, on a Proton
November 10	Russia publishes proposed constitution
December 2	U.S. launches STS-61, space shuttle *Endeavour*; astronauts successfully complete repairs of the Hubble Telescope
December 8	Yeltsin, meeting with NATO Secretary General, expresses disapproval of expanding NATO to include the Czech Republic, Poland, Slovakia or other East European nations
December 12	Russian elections: Yeltsin constitution narrowly approved; nationalist Zhirinovsky wins the highest percentage of party-preference vote (24%) in public rebuke of Yeltsin's reforms
December 16-17	U.S. and Russia officially agree to merge space station programs at third Gore-Chernomyrdin meeting
December 22	Russia launches *Molniya 1*, a primarily military comsat, on a Molniya booster
winter	4.2 million people are endangered in the war zone of former Yugoslavia[7]

1994

January 8	Russian launches *Soyuz TM-18*, sending 3 cosmonauts to Mir
January 10-11	NATO, in Brussels summit, announces limited "Partnership for Peace" plan for East European nations hoping to join the alliance

January 13-14   Clinton and Yeltsin meet in Moscow summit; Ukrainian
                 President Kravchuk joins them to sign a trilateral agreement
                 to turn Ukraine's nuclear warheads over to Russia for
                 destruction

January 17   Gaidar resigns from Yeltsin cabinet; reformist Finance Minister
                 Fyodorov follows on January 27, setting off a record fall in the
                 ruble and casting doubt on Russian economic reform

January 20   Russia launches *Gals 1*, 1st of new series of comsats on Proton

January 28   Russian launches *Progress M-21* to resupply *Mir*

February 3   U.S. launches STS-60, space shuttle *Discovery*, carrying
                 Krikalev, first Russian cosmonaut aboard a U.S. space mission
                 Georgia signs a military cooperation treaty with Russia,
                 seeking aid in civil war
                 Ukrainian Parliament renounces the conditions it had placed
                 on ratification of START 1, but avoids approving accession to
                 the Nuclear Non-Proliferation Treaty

February 5   Russia launches *Raduga-1*, a primarily military comsat

February 12   Russia launches *Cosmos 2268-2273*, six-pack of store-and-
                 forward comsats

February 14   Clinton announces U.S. will triple aid to Kazakhstan to $311
                 million

February 17   Russia unilaterally arranges a Serbian withdrawal from around
                 the besieged city of Sarajevo just on time to avoid a NATO
                 deadline

February 18   Russia launches *Raduga 31*, a primarily military comsat

February 22   CIA announces that its counterintelligence chief stationed in
                 Moscow has been a double agent for the last eight years,
                 setting off an exchange of diplomatic expulsions

February 28   Russia, supporting Serbs, announces intention to deploy 400
                 peacekeeping troops around Sarajevo

March 1   NATO jets shoot down 4 Serbian aircraft; NATO's first combat
                 action in its 45-year history

March 4   Clinton, in Washington meeting with Kravchuk, announces
                 U.S. will double aid to Ukraine to $700 million contingent
                 upon its parliament ratifying January 1994 agreement
                 U.S. launches STS-62, space shuttle *Columbia*

March 17   Russia launches *Cosmos 2274*, hi-res. film-rtrn. photo recon sat

April 9   U.S. launches STS-59

April 11   Russia launches *Cosmos 2275-2277*, part of the GLONASS
                 positioning constellation

April 23   Russia launches *Cosmos 2278*, an electronic intelligence satellite
                 (ELINT, 4th generation)

April 26   Russia launches *Cosmos 2279*, a navsat

April 28   Russia launches *Cosmos 2280*, a photo recon sat

May 3   U.S. launches a $1 billion NRO payload on a Titan 4

May 25   Russia fails in an attempt to launch an ELINT 3 satellite

June   NASA and RSA sign a Memorandum of Understanding

	outlining Shuttle-Mir goals
July 1	Russia launches *Soyuz TM-19* with two cosmonauts
June 7	Russia launches *Cosmos 2281*, a med.-res. photo recon sat
July 7	Russia launches *Cosmos 2282*, an advanced launch detection (LDS, or ballistic missile warning)
July 8	U.S. launches STS-65, space shuttle
July 14	Russia launches *Nadezhda 4*, a search and rescue satellite
July 20	Russia launches *Cosmos 2283*, a hi-res. photo recon sat
July 29	Russia launches *Cosmos 2284*, a photo recon sat
August 2	Russia launches *Cosmos 2285*, possibly a navsat
August 5	Russia launches *Cosmos 2286*, a launch detection satellite
August 11	Russia launches *Cosmos 2287-2289*, part of the GLONASS positioning constellation
August 23	Russia launches *Molniya 3*, a primarily military comsat
August 26	Russia launches *Cosmos 2290*, an new type of advanced photo recon sat
September 9	U.S. launches STS-64, space shuttle
September 21	Russia launches *Cosmos 2291*, a relay satellite
September 27	Russia launches *Cosmos 2292*, a calibration satellite
September 30	U.S. launches STS-68, space shuttle
October 3	Russia launches *Soyuz TM-20*; crew includes German
October 11	Russia launches *Okean 4*, for oceanographic ice mapping
October 13	Russia launches *Ekspress 1*, an advanced comsat
October 24	India joins the spacefaring nations by launching its first Polar Satellite launch vehicle
November	proposed U.S.-Russian Mars Together mission falls apart
November 2	Russia launches *Cosmos 2293*, an ELINT ocean recon sat
November 3	U.S. launches STS-66
November 20	Russia launches *Cosmos 2294-2296*, part of the GLONASS positioning constellation
November 24	Russia launches *Cosmos 2297*, an ELINT 4 on a Zenit booster
November 29	Russia launches *GEO-IK 1*, a geodetic satellite
December 14	Russia launches *Molniya 1* a primarily military comsat
December 20	Russia launches *Cosmos 2298*, a store-and-forward satellite
December 28	Russia launches *Cosmos 2299-2304*, a six-pack store-and-forward satellite
December 28	Russia launches a *Raduga* satellite
December 29	Russia launches *Cosmos 2305*, a photo recon sat

1995

February 3	U.S. launches Space Shuttle *Discovery*, carrying Titov, the second Russian cosmonaut aboard a U.S. space mission; the shuttle flies within 37 feet of Mir
March 2	Russia launches *Cosmos 2306*, a radar calibration satellite (for ABM use?)
March 3	U.S. launches space shuttle *Endeavour* with the Astro-2 ultraviolet telescope package

March 7	Russia launches *Cosmos 2307-2309*, part of the GLONASS positioning constellation
March 14	Russia launches *Soyuz TM-21*, with cosmonauts and U.S. astronaut Norman Thagard to Mir
March 22	Russia launches *Cosmos 2310*, a navsat
	Russia launches *Cosmos 2311*, a high-res. photo recon sat
May 24	Russia launches *Cosmos 2312*, an early-warning satellite
June 27	U.S. launches space shuttle *Atlantis*, which docks with Mir space station on June 29 (the first Shuttle-Mir docking). Together the two vehicles set a record for astronauts on one spacecraft (10).
July 13	U.S. launches space shuttle *Discovery*, after a 5-week delay caused by woodpecker holes in the external fuel tank
August 16	The first launch of the first entirely commercially developed booster, Lockheed Martin's LLV-1, explodes over the Pacific on its first launch attempt
October	U.S. to launch STS-74 for docking with Mir

1996

March	U.S. to launch STS-76 for docking with Mir
August	U.S. to launch STS-79 for docking with Mir
December	U.S. to launch STS-81 for docking with Mir

1997

May	U.S. to launch STS-84 for docking with Mir
September	U.S. to launch STS-86 for docking with Mir
November	Space station Alpha construction to begin with launch of FGB module, built by Khrunichev in direct contract from Boeing

Notes

1. *PP of JFK, 1961* (Washington: GPO, 1962) 2.

2. Ibid 26-27.

3. Michael Stoiko, *Soviet Rocketry* (New York: Holt, Rinehart and Winston, 1970) 110-11.

4. Wernher von Braun and Frederick I. Ordway, *Space Travel: A History* (New York: Harper and Row, 1985) 204.

5. Department of Defense, "The Soviet Space Challenge," (Washington: Office of the Secretary of Defense, Nov. 1987) 16.

6. This number was secret and may be incorrect, but the satellite's purpose is known.

7. Steven A. Holmes, "Next Worry: North Korea," *New York Times* 14 Nov. 1993: A18.

Bibliography

I. Books and Dissertations

Air War College, 1988 Space Issues Symposium. *Building a Consensus Toward Space*. Maxwell Air Force Base, Alabama: Air UP, Apr. 1990.

Allison, Graham T. *Essence of Decision*. Boston: Little, Brown and Company, 1971.

Amann, Ronald, and Julian Cooper. *Industrial Innovation in the Soviet Union*. New Haven: Yale UP, 1982.

------. *Technical Progress and Soviet Economic Development*. New York: Basil Blackwell, 1986.

Ambrose, Stephen E. *Ike's Spies: Eisenhower and the Espionage Establishment*. Garden City: Doubleday, 1981.

Avduevskiij, B. S. *M. V. Keldysh*. Moscow: Nauka, 1988.

Bainbridge, William Sims. *Goals in Space*. Albany: State University of New York Press, 1991.

Beschloss, Michael R. *The Crisis Years*. New York: Harper Collins, 1991.

Bilstein, Roger E. *Orders of Magnitude*. Washington: NASA, 1989.

Bluth, Christoph. *Soviet Strategic Arms Policy Before SALT*. Cambridge, Cambridge UP, 1992.

Brown, Neville. *New Strategy Through Space*. Leicester: Leicester UP, 1990.

Bulkeley, Rip and Graham Spinardi. *Space Weapons*. Totowa, New Jersey: Barnes and Noble Books, 1986.

Butterworth, Robert L. *Guide to Space Issues for the 1990s*. Los Alamos: Center for National Security Studies, 1992.

Byerly, Radford Jr. *Space Policy Alternatives*. Boulder: Westview Press, 1992.

------. *Space Policy Reconsidered*. Boulder: Westview Press, 1989.

Carter, Jimmy. *Keeping Faith: Memoirs of a President*. Toronto: Bantam Books, 1982.

Chagas, Carlos, and Vittorio Canuto. The Vatican: Pontificia Academia Scientiarvm, 1986.

Cline, Ray S. *The CIA Under Reagan, Bush and Casey* Washington: Acropolis Books, 1981.

------. *Space Support of U.S. National Security*. Conference Report. Washington: U.S. Global Strategy Council, 1988.

Clark, Phillip. *The Soviet Manned Space Program*. New York: Orion Books, 1988.

Collins, John M. *Military Space Forces*. Washington: Perhamon-Brassey's, 1989.

Collins, Martin J., and Sylvia D. Fries. *A Spacefaring Nation: Perspectives on American Space History and Policy*. Washington: Smithsonian Institution Press, 1991.

Collins, Michael. *Mission to Mars*. New York: Grove Weidenfeld, 1990.

The Commission for the Scientific Legacy of the Pioneers of the Mastery of Space. *Iz Istorii Sovestskoj Kosmonavtiki* (From the History of Soviet Cosmonauts). Moscow: Nauka, 1983.

Cook, Chris, and John Stevenson. *Longman Handbook of World History Since 1914*. New York: Longman, 1991.

Cozic, Charles P. *Space Exploration: Opposing Viewpoints*. San Diego: Greenhaven Press, 1992.

Crankshaw, Edward. *Khrushchev: A Career*. New York: Viking, 1966.

Divine, Robert A. *The Sputnik Challenge*. New York: Oxford UP, 1993.

Dunlop, John B. *The Rise of Russia and the Fall of the Soviet Empire*. Princeton: Princeton UP, 1993.

Dutton, Lyn. *Military Space*. vol. 10 of *Brassey's Air Power: Aircraft, Weapons Systems and Technology Series*. London: Brassey's, 1990.

Eisenhower, Dwight D. *Mandate for Change*. Garden City: Doubleday, 1963.

------. *Waging Peace*. Garden City: Doubleday, 1965.

El-Baz, Farouk, and D. M. Warner. *Apollo-Soyuz Test Project: Summary Science Report* vol. 2. Washington: NASA Scientific and Technical Information Branch, 1979.

Evangelista, Matthew. *Technical Innovation and the Arms Race*. Ithaca: Cornell UP, 1988.

Ezell, Edward Clinton, and Linda Neuman Ezell. *The Partnership, A History of the Apollo-Soyuz Test Project*. Washington: NASA, 1978.

Felshman, Neil. *Gorbachev, Yeltsin and the Last Days of the Soviet Empire*. New York: St. Martin's Press, 1992

Fleron, Frederic J., Jr., and Erik P. Hoffmann. *Post-Communist Studies and Political Science*. Boulder: Westview Press, 1993.

Fleron, Frederic J., Jr., Erik P. Hoffmann and Robbin F. Laird. *Contemporary Issues in Soviet Foreign Policy*. New York: Aldine de Gruyter, 1991.

Frutkin, Arnold W. *International Cooperation in Space*. Prentice Hall: Englewood Cliffs, 1965.

Gaddis, John Lewis. *Strategies of Containment: A Critical Appraisal of Postwar American National Security Policy*. New York: Oxford UP, 1982.

Garthoff, Raymond L. *Detente and Confrontation: American-Soviet Relations from Nixon to Reagan*. Washington: The Brookings Institution, 1985.

------. *Reflections on the Cuban Missile Crisis*. Washington: The Brookings Institution, 1989.

George, Alexander L., and Philip J. Farley and Alexander Dallin. *U.S.-Soviet Security Cooperation*. New York: Oxford UP, 1988.

Goldman, Marshall I. *What Went Wrong with Perestroika*. New York: W. W. Norton, 1991.

Goldman, Nathan C. *Space Policy*. Ames: Iowa State UP, 1992.

Grabbe, Crockett L. *Space Weapons and the Strategic Defense Initiative*. Ames: Iowa State UP, 1991.

Graham, Loren. *Science In Russia and the Soviet Union*. New York: Cambridge UP, 1993.

Gray, Colin S. *American Military Space Policy*. Cambridge: Abt Books, 1982.

Groman, Jeff. *NASA*. London: Multimedia Publications, 1986.

Gump, David P. *Space Enterprise*. New York: Praeger, 1990.

Harvey, Brian. *Race into Space*. New York: Halsted Press, 1988.

Harvey, Dodd L., and Linda C. Ciccoritti. *U.S.-Soviet Cooperation in Space*. Miami: Center for Advanced International Studies, 1974.

Holloway, David. *The Soviet Union and the Arms Race*. New Haven: Yale UP, 1983.

------. *Stalin and the Bomb*. New Haven: Yale UP, 1994.

Humble, Ronald D. *The Soviet Space Programme*. London: Routledge, 1988.

Hurt, Harry. *For All Mankind*. New York: The Atlantic Monthly Press, 1988.

Ishlinskiij, A. Yu. *Akademik S. P. Korolev* (Academic S. P. Korolev). Moscow: Nauka, 1986.

Jamgotch, Nish. *U.S.-Soviet Cooperation*. New York: Praeger, 1989.

Jasani, Bhupendra. *Outer Space: A New Dimension of the Arms Race*. London: Taylor and Francis, 1982.

------. *Space Weapons: The Arms Control Dilemma*. London: Taylor and Francis, 1984.

Johnson, Nicholas L. *Soviet Military Strategy in Space*. London: Jane's Publishing, 1987.

------. *The Soviet Year in Space* vols. 1985-1990. Colorado Springs: Teledyne Brown Engineering, 1986-1991.

------. *Europe and Asia in Space* vol. 1991-1992. Colorado Springs: Kaman Sciences Corp., 1994.

Kash, Don E. *The Politics of Space Cooperation*. West Lafayette: Purdue University Studies, 1967.

Kerrod, Robin. *The Illustrated History of NASA*. London: Prion, 1988.

Karpenko, A. V. *Rossiijskoe Raketnoe Oruzhie, 1943-1993 g.g.* (Russian Rocket Forces, 1943-1993). Saint Petersburg, PIKA, 1993.

Khrushchev, Nikita Sergeevich. *Khrushchev Remembers*. Trans. Strobe Talbott. Boston: Little, Brown and Company, 1970.

------. *Khrushchev Remembers: The Glasnost Tapes*. Trans. Jerry L. Schecter. Boston: Little, Brown and Company, 1990.

Khrushchev, Sergei. *Khrushchev on Khrushchev*. Trans. William Taubman. Boston: Little, Brown and Company, 1990.

------. *Krizisi i Rakety* (Crises and Rockets). 2 vols. Moscow: Novosti, 1994.

Kistiakowsky, George B. *A Scientist at the White House*. Cambridge: Harvard UP, 1976.

Klass, Philip J. *Secret Sentries in Space*. New York: Random House, 1971.

Krieger, F. K. *Behind the Sputniks*. Washington: Public Affairs Press, 1958.

Krug, Linda T. *Presidential Perspectives on Space*. New York: Praeger, 1991.

Legvold, Robert. "Foreign Policy." *After the Soviet Union*. Ed. Timothy J. Colton and Robert Legvold. New York: W. W. Norton, 1992. 147-176.

Lewis, Richard S. *Space in the 21st Century*. New York: Columbia UP, 1990.

Logsdon, John M. *The Decision to Go to the Moon*. Chicago: The University of Chicago Press, 1970.

Long, Franklin A., Donald Hafner and Jeffrey Boutwell. *Weapons in Space*. New York: W. W. Norton, 1986.

Luongo, Kenneth M. and W. Thomas Wander. *The Search for Security in Space.* Ithaca: Cornell UP, 1989.

MacDonald, Brian. *Space Strategy: Three Dimensions.* Toronto: Canadian Institute of Strategic Studies, 1989.

MacKinnon, Douglas, and Joseph Baldanza. *Footprints.* Washington: Acropolis Books, 1989.

McAuley, Mary. *Soviet Politics, 1917-1991.* Oxford: Oxford UP, 1992.

McCurdy, Howard E. *The Space Station Decision.* Baltimore: Johns Hopkins UP, 1990.

McDougall, Walter A. *The Heavens and the Earth.* New York: Basic Books, 1985.

McElroy, John H., and Brent Scowcroft. *A More Effective Civil Space Program.* Washington: Center for Strategic and International Studies, 1989.

Medvedev, Roy A. *Khrushchev.* Trans. Brian Pearce. Oxford: Basil Blackwood, 1982.

Medvedev, Roy A., and Zhores A. Medvedev. *Khrushchev: The Years in Power.* New York: W. W. Norton and Company, 1978.

Mehta, Dalpat Singh. *Mass Media in the USSR.* Moscow: Progress Publishers, 1987.

Michaud, Michael A. G. *Reaching for the High Frontier: The American Pro-Space Movement, 1972-84.* New York: Praeger, 1986.

Mozhorin, Ju. A. *Kosmonavtika SSSR.* Moscow: Mashinostroenie, 1981.

------. *Kosmonavtika SSSR.* Moscow: Mashinostroenie, 1986.

Murray, Bruce. *Journey Into Space.* New York: W. W. Norton, 1989.

Muske, Edmund S. *The U.S. in Space.* Washington: Center for National Policy Press, 1988.

National Commission on Space, Report of the. *Pioneering the Space Frontier.* Toronto: Bantam Books, 1986.

Newkirk, Dennis. *Almanac of Soviet Manned Space Flight.* Houston: Gulf Publishing Company, 1990.

Newton, David. *U.S. and Soviet Space Programs.* New York: Franklin Watts, 1988.

Oberg, James E. *The New Race for Space.* Harrisburg: Stackpole Books, 1984.

------. *Red Star in Orbit.* New York: Random House, 1981.

Ordway, Frederick I., and Randy Liebermann. *Blueprint for Space.* Washington: Smithsonian Institution Press, 1992.

Osmanczyk, Edmund Jan. *The Encyclopedia of the United Nations and International Agreements.* Philadelphia: Taylor and Francis, 1985.

Papp, Daniel S., and John R. McIntyre. *International Space Policy.* New York: Quorum Books, 1987.

Pardoe, Geoffrey K. C. *The Future of Space Technology.* London: Francis Pinter, 1984.

Petrov, G. I. *Conquest of Outer Space in the USSR.* Moscow: Nauka Publishers, 1971.

Pomeroy, John J., and Norman J. Hubbard. *The Soviet-American Conference on Cosmochemistry of the Moon and Planets* pt. 2. Washington: NASA Scientific and Technical Information Office, 1977.

Richelson, Jeffrey T. *America's Secret Eyes in Space.* New York: Harper and Row, 1990.

Rokke, Ervin Jerome. *Politics of Aerial Reconnaissance: Eisenhower Administration.* Diss. Harvard University, 1971. Cambridge: privately printed, 1971.

Russian Strategic Rocket Forces. *Raketnye Voijska Strategicheskovo Naznachenija* (Strategic Rocket Forces). Moscow: Strategic Rocket Forces, 1992.

------. *Xronika Osnovyx Sobytiij Istorii Raketnyx Voick Strategicheskovo Naznachenija* (Chronicle of the Fundamental Events of the History of the Strategic Rocket Forces). Moscow: Strategic Rocket Forces, 1994.

Schaerf, Carlo, Guiseppi Longo and David Carlton. *Space and Nuclear Weaponry in the 1990s.* London: Macmillan, 1992.

Schauer, William H. *The Politics of Space: a Comparison of the Soviet and American Space Programs.* New York: Holmes and Meier, 1976.

Scott, Harriet Fast. *Soviet Military Strategy.* New York: Crane and Russak, 1975.

Shaffer, Stephen M., and Lisa Robock Shaffer. *The Politics of International Cooperation: A Comparison of U.S. Experience in Space and in Security.* Denver: University of Denver Graduate School of International Studies, 1980.

Smith, Delbert D. *Communications via Satellite: A Vision in Retrospect.* Boston: A. W. Sijthoff, 1976.

Smolders, Peter L. *Soviets in Space.* Trans. Marian Powell. New York: Taplinger, 1971.

Sokol'sky, Viktor N. *A Short Outline of the Development of Rocket Research in the U.S.S.R.* Jerusalem: Israeli Scientific Translations, 1967.

Stares, Paul B. *The Militarization of Space: U.S. Policy, 1945-1984.* Ithaca: Cornell UP, 1985.

------. *Space and National Security.* Washington: Brookings Institution, 1987.

------. *Space Weapons and U.S. Strategy: Origins and Development.* London: Croom Helm, 1985.

Stockholm International Peace Research Institute. *SIPRI Yearbook* 1992, 1993 and 1994. Oxford: Oxford UP, 1992, 1993 and 1994.

Stoiko, Michael. *Soviet Rocketry.* New York: Holt, Rinehart and Winston, 1970.

Tarasenko, M. V. *Voennye Aspekty Sovetskoj Kosmonavtiki.* Moscow: Nikol' Agenstvo Rossijskoj Pechati, 1992.

Taylor, L. B. *For All Mankind.* New York: E. P. Dutton, 1974.

Thompson, Wayne C., and Steven W. Guerrier. *Space: National Programs and International Cooperation.* Boulder: Westview Press, 1989.

Tolstoy, Leonid Nikolaevich. *War and Peace.* Trans. Rosemary Edmonds. London: Penguin Books, 1978.

Trager, Oliver. *America in Space: Pioneers or Aggressors?* New York: Facts on File, 1986.

Van Dyke, Vernon. *Pride and Power.* Urbana: University of Illinois Press, 1964.

Vladimirov, Leonid. *The Russian Space Bluff.* Trans. David Floyd. New York: The Dial Press, 1973.

von Borcke, Astrid. "Gorbachev's Perestroika: Can the Soviet System be Reformed?" *Gorbachev's Agenda.* Ed. Susan L. Clark. Boulder: Westview Press, 1989. 13-56.

von Braun, Wernher, and Frederick I. Ordway. *Space Travel: A History.* New York: Harper and Row, 1985.

Wädekin, Karl-Eugen. "Agriculture." *Gorbachev and Perestroika.* Ed. Martin

McCauley. London: Macmillan, 1990. 70-95.

Webb, James E. *Space Age Management*. New York: McGraw-Hill, 1969.

Weber, Steve. *Cooperation and Discord in U.S.-Soviet Arms Control*. Princeton: Princeton UP, 1991.

White, Stephen. *Gorbachev and After*. Cambridge: Cambridge UP, 1992.

Wilson, Andrew. *Jane's Space Directory* 1994-95. Coulsdon: Jane's Data Division, 1994.

Wohlstetter, Roberta. *Pearl Harbor: Warning and Decision*. Stanford: Stanford UP, 1962.

Yeltsin, Boris. *The Struggle For Russia*. New York: Random House, 1994.

York, Herbert F. *Arms Control: Readings from Scientific American*. San Francisco: W. H. Freeman, 1973.

II. Periodicals, Articles and Other

Abramova, Yelena "International Center to Finance Science in Russia." Moscow *TASS* 30 Jun. 1995.

"Academician Discusses U.S.-Soviet Space Cooperation Plans." Moscow *Pravda* 27 May 1977. Translation by the Foreign Broadcast Information Service. *FBIS Daily Report--Soviet Union* vol. 3: 109; 7 Jun. 1977: 3-4. Hereafter referred to as *FBIS*.

"Accident Reveals 'Disastrous' Conditions at Baikonur Town." Moscow *TASS* 28 Jul. 1994. *FBIS* 29 Jul. 1994: 33.

"Agreement on Guidelines for the Transfer of Equipment and Technology Related to Missiles." *International Legal Materials* vol. 26, no. 3, May 1987: 599-613.

"Agreements Establishing the Commonwealth of Independent States." *International Legal Materials* vol. 31, no. 1, Jan. 1992: 138-154.

Aldhous, Peter. "Can Russia Slim Down to Survive?" *Science* 19 Nov. 1993: 1200-1202.

------. "New Foundation Goes Back to Basics." *Science* 19 Nov. 1993: 1200.

Alexander, Bretton S., Tom Cremins and James Oberg. "Exchange: The State of U.S.-Russian Space Cooperation." *Washington Times* 6 Jan. 1994: A19.

"American Physical Society and American Astronautical Society Launch Aid Programs for Former Soviet Union Scientists." *Physics Today* May 1992: 56.

Anderson, Christopher. "Russian Science Aid Falls Short." *Science* 10 Sep. 1991: 1380.

Aouizerate, Paul. "An Open Letter to the President of Globalstar," *Space News* 5-11 Jun. 1995: 5.

Asker, James R. "Clinton Budget Cuts NASA 'To the Bone.'" *Aviation Week and Space Technology* 14 Feb. 1994: 26. Hereafter referred to as *Aviation Week*.

------. "Commercial Growth Key to Space Sector." *Aviation Week* 13 Mar. 1995: 95-97.

------. "Cosmonaut Upstages Shuttle Payloads." *Aviation Week* 31 Jan. 1994: 58-61.

------. "Mars 94 to Carry Novel U.S. Experiment." *Aviation Week* 12 Apr. 1993: 53 and 56-57.

------. "Nearly 1,000 Spacecraft." *Aviation Week* 16 Jan. 1995: 55.

------. "Racing to Remake Space Economics." *Aviation Week* 3 Apr. 1995: 45-48, 53.

------. "Soviet Space Programs Bid for Funding in Weak Economy." *Aviation Week* 6 May 1991: 23.
------. "Space Key to U.S. Defense." *Aviation Week* 3 May 1993: 57.
------. "Space Station Key to NASA's Future." *Aviation Week* 15 Mar. 1993: 81-83.
------. "U.S. Firm Says Baikonur Improving Slowly." *Aviation Week* 31 Jul. 1995: 25-26.
Aslund, Anders. "Ruble Awakening." *Washington Post* 23 Apr. 1995: C4.
"Average Russian Income Drops." *OMRI Daily Digest* no. 95, pt. 1, 17 May 1995.
Baev, Andrei, Matthew J. Von Bencke, David Bernstein, Jeffrey Lehrer and Elaine Naugle. "American Ventures in Russia." Stanford: Center for International Security and Arms Control, 1995.
Baberdin, Valeriy. "Does Russia Need Baikonur?" *Krasnaya Zvezda* (22 Oct. 1994: 4) FBIS 25 Oct. 1994: 26-27.
"Baikonur Chief Blasts U.S. Government Study." *Space News* 14-20 Feb. 1994: 2.
"Baikonur 'Will Not Become a Russian Military Base.'" *Moscow Ostankino Television First Channel Network* 29 Dec. 1994. FBIS 10 Jan. 1995: 5.
"Baikonur 'Will Not Be Needed' In a Few Years." *Moscow Ostankino Television First Channel Network* 28 Aug. 1994. FBIS 31 Aug. 1994: 36-37.
"Battle Expands Over Shrinking Budget." *Science* 14 Jan. 1994: 166.
Beardsley, Tim. "Brain Drain." *Scientific American* Apr. 1992: 17-18.
------. "Selling to Survive." *Scientific American* Feb. 1993: 92-100.
Bohlen, Celestine. "For Russia and Ukraine, Crimea is a Sore Nerve." *New York Times* 23 Mar. 1994: A3.
------. "Moscow Rumors, True or Not, Show The Weakening of Yeltsin's Position." *New York Times* 27 Mar. 1994: A12.
------. "Russian Economy Has I.M.F. Checkup." *New York Times* 18 Mar. 1994: A9.
Broad, William J. "Daniel S. Goldin: Bold Remodeler of the Drifting Space Agency." *New York Times* 21 Dec. 1993: C1.
------. "Genius for Hire: The Soviet's Best, at Bargain Rates." *New York Times* 15 Mar. 1992: E3.
Browne, Malcolm W. "End of Cold War Clouds Research As Openings in Science Dwindle." *New York Times* 20 Feb. 1994: A1.
Budiansky, Stephen. "A Scientific Bazaar." *U.S. News and World Report* 4 May 1992: 58-60.
Burt, Richard. "Soviet Said to Ask for Space Shuttle Halt." *New York Times* 1 Jun. 1979: A6.
"But Tax Collections Improve." *Monitor* vol. 1, no. 42, 29 Jun. 1995.
Charles, Dan. "Bargain Hunters Snap Up Russian Brainpower." *New Scientist* 14 Mar. 1992: 13.
"Chernomyrdin Describes Economy." *OMRI Daily Digest* no. 140, pt. 1, 20 Jul. 1995.
Clarke, Doug. "Hard Times in Defense Industry." *RFE/RL Daily Report* no. 221, 22 Nov. 1994.
------. "U.S. Companies to Market Russian Military Satellite Images." *OMRI* no. 145, pt. 1, 27 Jul. 1995.
Cohen, Roger. "NATO and Russia Clash on Future Alliances." *New York Times* 10 Dec. 1993: A15.

Covault, Craig. "Mars Exploration Invites Global Space Cooperation." *Aviation Week* 12 Apr. 1993: 56-57.

------. "Mars Strategy Begs for Direction." *Aviation Week* 5 Oct. 1992: 25-26.

------. "Mobilize 'Space Year' to Save Russian Mars 94 Flight." *Aviation Week* 13 Jan. 1993: 35-36.

------. "NASA, Soviets Discuss Joint Environmental Missions." *Aviation Week and Space Technology* 7 Oct. 1991: 68-69.

------. "Russia/CIS Space Outlook Chaotic But Critical to Global Planning." *Aviation Week* 16 Mar. 1992: 125.

------. "Russia Forges Ahead on Mir 2." *Aviation Week* 15 Mar. 1993: 26-27.

------. "Russia Launches Three Spy Satellites." *Aviation Week* 27 Sep. 1993: 24.

------. "Russian Military Space Program Maintains Aggressive Pace." *Aviation Week* 3 May 1993: 61.

------. "Russians Locked in Struggle for Space Program Control." *Aviation Week* 1 Feb. 1992: 57-59.

------. "U.S., Russia Plan New Mars Mission." *Aviation Week* 6 Jun. 1994: 24.

Cowen, Robert C. "East-West Science--the Uneasy Detente." *Christian Science Monitor* 2 Jul. 1975: 21.

"Crimea Will Hold a Runoff: Rivals Vie on Region's Fate." *New York Times* 18 Jan. 1994: A10.

Crossette, Barbara. "Russia's Rocket Deal With India Leads U.S. to Impose Trade Bans." *New York Times* 12 May 1992: A8.

David, Leonard. "Boosters versus Bread Lines." *Ad Astra* Jul./Aug. 1992: 6.

David, Leonard, and Joanne Padron. "International Space Cooperation." Washington: American Institute of Aeronautics and Astronautics, Mar. 1993.

"Russia Considers Accompanying U.S. in Pluto Exploration." *Space News* 2-8 May 1994: 14.

------. "Electrifying: Soviet Show and Tell." *Ad Astra* Dec. 1991: 5-6.

------. "U.S.-Russian Exchange Program." *Ad Astra* Sep./Oct. 1992: 6.

"DC-X Demonstrates Key Maneuver." *Aviation Week* 17 Jul. 1995: 28.

"Decline in Russian GDP Widened." *Wall Street Journal* 18 Jan. 1995: A10.

"DEC Exploits Russian Aerospace Niche." *Aviation Week* 28 Jun. 1993.

Delpech, Jean-Francois, John M. Logsdon and Brad M. Meslin. *International Cooperation in Space: Strategies for the New Century*. Arlington: Center for Research and Education on Strategy and Technology, May 1993.

"Device is Eight Times Heavier Than One Planned by U.S." *New York Times* 5 Oct. 1957: A1.

Dornheim, Michael A. "Aerojet Tests Russian Oxidizer-Cooled Thruster." *Aviation Week* 27 Jun. 1994: 75.

------. "DC-X Resumes Flying After Hiatus." *Aviation Week* 27 Jun.1994: 26.

------. "Russians Press Mars 94, But Risks Still Run High." *Aviation Week* 3 Jan. 1994: 22-23.

Dudkin, Lev, and Anatol Vasilevsky. "The Soviet Military Burden: A Critical Analysis of Current Research." *Hitotsubashi Journal of Economics* vol. 28, 1987: 41-61.

"Duma Deputies Term Baikonur Situation 'Disastrous.'" Moscow *TASS* 13 Jul. 199. *FBIS*. 15 Jul. 1994: 22.

"Duma Ratifies Treaty on Renting Baikonur." Moscow *TASS* 21 Apr. 1995. *FBIS* 24 Apr. 1995: 32.

"East/West Aerospace Ventures on the Rise." *Aviation Week* 23 Aug. 1993: 51-52.

"Ekonomicheskaya Situatziya v Rossii." *Ekonomika i Zhizn'* no. 1, Jan. 1995: 1.

Erlanger, Steven. "Accepting Reformer's Resignation, Yeltsin Tries to Prevent Another." *New York Times* 18 Jan. 1994: A8.

------. "Amnesty of Foes Brings Disarray to Yeltsin Team." *New York Times* 28 Feb. 1994: A1.

------. "Finance Minister Shuns Yeltsin Plea and Quits Cabinet." *New York Times* 27 Jan. 1994: A1.

------. "Inflation and Unpaid Bills Haunt Russia as it Works on '94 Budget." *New York Times* 14 Feb. 1994: A9.

------. "Reformer Quits New Yeltsin Cabinet." *New York Times* 21 Jan. 1994: A1.

------. "Rightist Vote Helps Russia Fight Expansion of NATO." *New York Times* 29 Dec. 1993: A1.

------. "Ruble Sinks Further in a Whirl of Political Rumors." *New York Times* 20 Jan. 1994: A3.

------. "Russian Premier's Star is Rising Fast." *New York Times* 26 Jan. 1994: A6.

------. "Russia's Reform Train Isn't Exactly an Express." *New York Times* 23 Jan. 1994: A5.

"Excerpts From President Clinton's Message on the State of the Union." *New York Times* 26 Jan. 1994: A16.

Feinberg, Evgenii L. "Soviet Science in Danger." *Physics Today* May 1992: 30-37.

Ferster, Warren. "Launch Firms Fight Foreign Missile Use." *Space News* 6-12 Mar. 1995: 1, 20.

------. "Russia Did Not Violate MTCR, Official Says." *Space News* 12-18 Jun. 1995: 12.

------. "Russia: Relax Launch Limits." *Space News* 19-25 Dec. 1994: 1, 21.

------. "U.S. Firms Lock Horns Over Launch Quotas." *Space News* 12-18 Jun. 1995: 1, 37.

"The Final Tally." *The Economist* 8 Jan. 1994: 55.

Friedman, Louis. "Return to the Martian Surface." *Ad Astra* Sep./Oct. 1992: 29-32.

Friedman, Thomas L. "One Topic, Several Agendas and Clinton and Nixon Meet." *New York Times* 9 Mar. 1993: A1.

------. "Russia Policy: A U.S. Riddle." *New York Times* 27 Jan. 1994: A1.

------. "U.S. Asks Allies to Help Speed I.M.F. Aid to Russia." *New York Times* 1 Feb. 1994: A6.

------. "U.S. to Offer Plan to Keep Scientists at Work in Russia." *New York Times* 8 Feb. 1992: A1.

"Future of Space Complex, Program Reviewed." Moscow *Izvestia* 4 Oct. 1991. *FBIS* vol. 3: 198; 11 Oct. 1991: 2-3.

Fyodorov, Boris. "Moscow Without Mirrors." *New York Times* 1 Apr. 1994: A27.

Gande, William. "Rovers Tested for International Mars Mission." *Ad Astra*
Sep./Oct. 1992: 16.

Gardner, Richard N. "Cooperation in Outer Space." *Foreign Affairs* vol. 41, no. 2.
Jan. 1963: 344-59.

Gavaghan, Helen. "Soviets Search for Partners in Space." *New Scientist* 2 Nov.
1991:17-18.

Gertz, Bill. "Russia Sells Rocket Motors to China." *Washington Times* 13 Feb. 1995:
A4.

"Glasnost in Space." *Science News* 7 Sep. 1991: 156.

"Government Approves 1996 Draft Budget." *OMRI Daily Digest* no. 151, pt. 1, 4
Aug. 1995.

"Government Decree on Baikonur Space Center," *Rossiyskaya Gazeta* 7 Sep. 1994:
4. *FBIS* 8 Sep. 1994: 29-30.

Graham, Loren. "Is Russian Science Dead?" Harvard Russian Research Center,
Director's Seminar, 17 Nov. 1993.

------. "Is Russian Science Dead?" Stanford University, Donald M. Kendall
Lectures, 7 Apr. 1995.

Gray, Malcolm. "The Last Soviet." *MacLean's* 24 Feb. 1992: 26-27.

Greenberg, D. S. "Space Notes: Soviet Guests; Restrictions on Military
Developments; West Ford." *Science* 20 Apr. 1962: 247.

Greenhouse, Steven. "Christopher Defends Russia Policy in Wake of CIA Agent's
Arrest." *New York Times* 3 Mar. 1994: A6.

------. "Clinton Vows to Improve Relations With Ukraine." *New York Times* 5 Mar.
1994: A5.

------. "Seven Offer Moscow Technical Help." *New York Times* 18 Jul. 1991: A1.

------. "U.S. to Russia: A Tougher Tone and a Shifting Glance." *New York Times*
21 Mar. 1994: A9.

"GDP Down But Some Sectors Show Growth." *OMRI Daily Digest* no. 135, pt. 1,
13 Jul. 1995.

Guttsman, Janet. "Russian Economy Seen Contracting." *Reuter European Business
Report* 24 Oct. 1994.

Hall, R. Cargill. "The Origins of U.S. Space Policy: Eisenhower, Open Skies, and
Freedom of Space." *Colloquy* Dec. 1993.

Harwood, William. "Pratt & Whitney Touts RD-120 Rocket Engine." *Space News*
31-6 Jul./Aug. 1995: 8.

------. "Tight Budget May Limit U.S. Trips to Mir." *Space News* 17-23 Jan. 1994:
19.

Hazard, J. N. "The Soviet Union and International Law." *Soviet Studies*, vol. 1, no.
3, Jan. 1950.

Hazarika, Sanjoy. "Despite U.S., Yeltsin Backs Rocket Deal With India." *New York
Times* 30 Jan. 1993: A2.

"Helping Soviet Science (continued)." *Nature* 19 Mar. 1992: 179-80.

"The High Price of Freeing Markets." *The Economist* 19 Feb. 1994: 57.

Holmes, Steven A. "Next Worry: North Korea." *New York Times* 14 Nov. 1993:
A18.

Iannotta, Ben. "Delegation Reports Baikonur Cosmodrome in Good State." *Space
News* 4-10 Aug. 1994: 12.

------. "JPL's Mars Pathfinder Program Taking Shape." *Space News* 19-25 Dec. 1994: 9.

------. "Officials Strive to Keep Mars Together Alive." *Space News* 17-23 Apr. 1995: 1, 20.

------. "Russia Accused of Underbidding to Win Greensat Launch." *Space News* 28-6 Feb./Mar. 1994: 8.

------. "U.S., Russia Learn as They Go." *Space News* 21-4 Nov./Dec. 1994: 1, 28.

------. "U.S. Spy Satellite Spending Static in 1995." *Space News* 14-20 Feb. 1994: 4.

"If Elections Were Held." *Monitor* vol. 1, no. 40, 27 Jun. 1995.

Jehl, Douglas. "Ukraine: A Nuclear Power, but Untested Loyalties." *New York Times* 2 Dec. 1993: A1.

Johnston, David. "Ex-Branch Leader of CIA is Charged as a Russian Agent." *New York Times* 23 Feb. 1994: A1.

"Joint U.S.-Russian Missions Planned." *Facts on File* 27 Aug. 1992: 628.

"Joint Venture Upgrades Baikonur Facilities." *Segodnya* 9 Aug. 1994: 3. *FBIS* 9 Aug. 1994: 41.

Julian, Thomas A. "Operations at the Margin: Soviet Bases and Shuttle-Bombing." *The Journal of Military History* Oct. 1993: 640.

Kanin, Yuri. "Exodus Likely to Increase." *Nature* 5 Dec. 1991: 342-43.

Kennedy, John F. "An Interview with John F. Kennedy." *Bulletin of the Atomic Scientists* vol. 16, no. 9. Nov. 1960: 346-47.

------. "If the Soviets Control Space, They Can Control the Earth." *Missiles and Rockets* 10 Oct. 1960: 11-12.

Kiernan, Vincent. "Soviets Cancel Fall Flight to Mir Station." *Space News* 29 Jul. 1991: 3.

------. "Russia Plans New Cosmodrome Outside Kazakhstan." *Space News* 21-27 Mar. 1994: 8.

Kifner, John. "Serbs Withdraw in a Russian Plan to Avert Bombing." *New York Times* 18 Feb. 1994: A1.

Kislov, A., and S. Krilov. "State Sovereignty in Air Space." *International Affairs* (Moscow) no. 3, Mar. 1956: 35-44.

Korovin, E. A. "Aerial Espionage and International Law." *International Affairs* (Moscow) no. 6, Jun. 1960: 49-50.

------. "International Status of Cosmic Space." *International Affairs* (Moscow) no. 1, Jan. 1959: 53-59.

Kozyrev, Andrei. "Don't Threaten Us." *New York Times* 18 Mar. 1994: A29.

"Lack of Funds Threatens Future Space Launches." Moscow *TASS* 14 Mar. 1995. *FBIS* 15 Mar. 1995: 29-30.

Lancaster, John. "Russia's Economic Prospects Fade, CIA Director Tells Hill." *New York Times* 26 Jan. 1994: A2.

Lavitt, Michael O. "New U.S./Russian Agreement." *Aviation Week* 12 Oct. 1992: 13.

------. "Phones for Tatarstan." *Aviation Week* 31 Aug. 1992: 13.

"Law Ratifies Baikonur Utilization Agreement," *Rossiyskaya Gazeta* 27 Oct. 1994: 6. *FBIS* 31 Oct. 1994: 32-33.

Lawler, Andrew. "East-West Launch Vehicle Venture in Turmoil." *Space News* 8-4 Aug. 1994: 5.

------. "U.S. To Begin Launch Talks With China, Russia." *Space News* 12-18 Sep. 1994: 1, 20.

"Learning From Russia's Reverse." Editorial. *New York Times* 29 Jan. 1994: A14.

Leary, Warren E. "Shuttle Lifts Off on First Mission With a Russian." *New York Times* 4 Feb. 1994: A17.

"Lebed Gains, Yeltsin Loses in Presidential Preference Poll." *Monitor* vol. 1, no. 53, 17 Jul. 1995.

Lemonick, Michael D. "Space Program for Sale." *Time* 16 Mar. 1992: 54-55.

Lenorovitz, Jeffrey M. "Lockheed, Khrunichev to Market Proton Launcher." *Aviation Week* 4 Jan. 1993: 24-25.

------. "Proven Technology is Cornerstone of Mir 2." *Aviation Week* 23 Aug. 1993: 60-62.

------. "Russian Maker of SLBMs Seeks Civilian Spin-offs." *Aviation Week* 9 Aug. 1993: 48-49.

------. "Russian Proton Booster Offered in Indonesian Launch Competition." *Aviation Week* 12 Apr. 1993: 61-62.

------. "Space Systems/Loral Books Proton Launches." *Aviation Week* 20 Sep. 1993: 90-91.

------. "U.S. Entrepreneurs Seek Russian SLBMs." *Aviation Week* 19 Apr. 1993: 22-23.

Leskov, Sergei. "Notes From a Dying Spaceport." *Bulletin of the Atomic Scientists* Oct. 1993: 40-44.

"Life During Reform: A Portrait in Numbers." *New York Times.* 16 Jun. 1996: A6.

Li, Vladimir. "Baikonur, Leninsk Difficulties Evaluated." *Kazakhstanskaya Pravda* 5 Jul. 1994. *FBIS* 9 Aug. 1994: 31-32.

"Lockheed in Space Venture with Russians." *New York Times* 29 Dec. 1992: B2.

"LKEI Board Approves Baikonur Upgrade Plan." *Business Wire* 6 Dec. 1994.

"LKEI Planning Four Proton Launches in 1996." *Aerospace Daily* 4 May 1995: 191.

"Loral Unit's Deal." *New York Times* 2 Apr. 1992: D2.

Maddox, John. "Russian Summer Hopes Dwarf Doubts." *Nature* 4 Jun. 1992: 357.

Maloney, Jim. "NASA Official Gets 'a little mad' Over Joint Space Mission Claims." *The Houston Post* 7 Nov. 1977: A7.

Mann, Paul. "U.S. Wins Accord on Kiev Warheads." *Aviation Week* 17 Jan. 1994: 22-23.

------. "Washington Outlook." *Aviation Week* 3 Oct. 1994: 25.

------. "Who'da Thunk It?" *Aviation Week* 26 Jun. 1995: 19.

Markoff, John. "Russia Computer Scientists Hired By American Company." *New York Times* 3 Mar. 1992: A1.

"Mars Rovers on the Move." *Sky and Telescope* May 1992: 487-88.

"Mass Misjudgment." *The Economist* 9 Oct. 1993: 57.

McKenna, James T. "Atlas Managers Await RD-180 Model." *Aviation Week* 12 Sep. 1994: 55.

Mellow, Craig. "Red Rocket's Glare." *Forbes* 26 Oct. 1992: 64-65.

Mervis, Jeffrey. "A Way to Wean Weapons Makers." *Nature* 27 Feb. 1992: 756.

------. "The West Gropes for Ways to Help." *Nature* 30 Apr. 1992: 733.

Mihalka, Michael. "Davydov on Foreign Debts, Investments." *OMRI Daily Digest* no. 123, pt. 1, 26 Jun. 1995.

"Mikhail Sergeevich Gorbachev is Elected President of the USSR." *Vremya* (television program). Moscow, 15 Mar. 1990.

"Military Industrial Complex in Trouble." *Monitor* vol. 1, no. 47, 10 Jul. 1995.

"Modified Russian Rover Operated By Satellite Link." *Aviation Week* 27 Feb. 1995: 61.

"More Russians Out on Strikes." *Monitor* vol. 1, no. 32, 14 Jun. 1995.

Morvant, Penny, and Thomas Sigel. "Concern About Brain Drain." *OMRI Daily Digest* no. 90, pt. 1, 10 May 1995.

Morvant, Penny. "Deterioration in Social Situation." *OMRI Daily Digest* no. 92, pt. 1, 12 May 1995.

------. "Russia Had Third Highest Suicide Rate." *OMRI Daily Digest* no. 100, pt. 1, 7 Jun. 1995.

"NASA, Russia, Partners Juggling Mars Missions." *Space News* 17-23 Apr. 1995: 20.

Newkirk, Dennis. "Cosmodrome News." *Countdown* Sep./Oct. 1994: 26-27.

Nikishin, Leonard. "Russia Has Given the Astro-Vehicle to America." *Obshchaya Gazeta* 8 Apr. 1994: 8. As translated by *JPRS*. "Alpha Station Said to Service Only U.S. Interests." *JPRS* 16 May 1994: 28-30.

Nixon, Richard. "Clinton's Greatest Challenge." *New York Times* 5 Mar. 1993: A21.

Oettinger, A. G. "An Essay in Information Retrieval or The Birth of a Myth." *Information and Control* vol. 8, no. 1. Cambridge: Academic Press, 1965.

"Official Says Russian Launch Agreement Could Be Revisited." *Aerospace Daily* 14 Feb. 1995: 233.

"Official Says 16,000 Servicemen to Staff Baikonur." Moscow *RIA* 27 Dec. 1994. *FBIS* 29 Dec. 1994: 4.

"One Russian in Ten." *Monitor* vol. 1, no. 28, 28 Jun. 1995.

"Output Falling More Slowly." *Monitor* vol. 1, no. 49, 11 Jul. 1995.

Pakhomov, Aleksandr. "Russia to Help NASA Launch Probes to Pluto, Sun." Moscow *TASS* 11 May 1994. *FBIS* 11 May 1994: 19.

"Parliament Adopts Budget, Ratified Baikonur Agreement." *Almaty Kazakh Radio Network* 14 Jul. 1994. *FBIS* 15 Jul. 1994: 52.

"Pluto Flight Considered." *Aviation Week* 9 May 1994: 27.

Possehl, Suzanne. "Its Budget Slashed, Russian Seed Bank Fights for Its Life." *New York Times* 23 Mar. 1993: C4.

Rebrov, Mikhail. "Mirages in the Visibility Zone, or In the Space Market Without Changes." *Krasnaya Zvezda*. As translated by FBIS: "Space Services Market Called Unfair." *FBIS* 24. Feb 1995: 1-3.

"Recipe for ex-Soviet Republics' Science." *Nature* 2 Jan 1992: 2.

"Retail Prices Increase in May." *OMRI Daily Digest* no. 100, pt. 1, 7 Jun. 1995.

Robinson, Leif J. "Tough Times for Astronomers in the Former Soviet Union." *Sky and Telescope* Sep. 1992: 254-58.

"Rockwell Agrees to Space Venture With Russians." *New York Times* 9 Sep. 1992: B2.

"Rockwell's Space Systems Division." *Aviation Week* 2 Nov. 1992: 19.

Rogachev, Vladimir. "Space Project With U.S. to Reach Sun, Pluto." Moscow *TASS* 11 Apr. 1994. *FBIS* 13 Apr. 1994: 30-31.

Rosett, Claudia. "Ruble's Slide Spotlights Power of Old-Style Central Bank Chief." *Wall Street Journal* 20 Jan. 1994: 1.

"Russia Cutting Fuel to Neighbors." *New York Times* 4 Mar. 1994: A1.

"Russia in Need." *The Economist* 15 Jan. 1994: 16-17.

"Russia Launches Boeing Experiment to Mir." *Aviation Week* 18 Oct. 1993: 30.

"Russia Leads CIS In Inflation." *Monitor* vol. 1, no. 63, 31 Jul. 1995.

"Russia Staking Its Claim in Commercial Launch Market." *Aerospace Daily* 28 Dec. 1994: 425.

"Russia, Ukraine Compete Over Australian Space Complexes," Moscow *TASS* 23 Feb. 1995. *FBIS* 23 Feb. 1995: 1.

"Russia Will Likely Import More Grain This Year." *Monitor* vol. 1, no. 51, 13 Jul. 1995.

"Russian Expert Says U.S. to Lift Satellite Quotas." *Reuters* 30 Jun. 1995.

"Russian Proton Challenges Ariane." *Aviation Week* 24 Apr. 1995: 40.

"Russian Space Program Seeks Needed Funding." *Aviation Week* 23 Aug. 1993: 24.

"The Russians May Have Succeeded in Sheltering Artifacts." *Space News* 3-9 Jan. 1994: 4.

"Russians Offer Use of Mir Space Station as a Prelude to Mars." *New York Times* 23 Feb. 1992: A26.

Sale, Murray. "Closing the File on Flight 007." *The New Yorker* 13 Dec. 1993: 90-101.

Sansom, Stephanie. "Looking for Alien Life." *Nature* 22 Aug. 1991: 653.

"Satellites and Our Safety." *Newsweek* 21 Oct. 1957: 29-39.

Schmemann, Serge. "On Russian TV, Clinton Backs Reforms." *New York Times* 15 Jan. 1994: A1.

------. "Russia Lurches Into Reform, But Old Ways Are Tenacious." *New York Times* 20 Feb. 1994: A1.

"Scientist Tells of USSR-U.S. Cooperation in Space Study." Moscow *Moscow Domestic Service* 30 May 1972. *FBIS* vol. 3: 106; 31 May 1972: 2-3.

"Scifor Corp." *PR Newswire* 16 Jun. 1994.

Sciolino, Elaine. "Move by Russians a Surprise to U.S." *New York Times* 18 Feb. 1994: A6.

Scott, William B. "Martin Says RD-180/Atlas Looks Promising." *Aviation Week* 5 Dec. 1994: 50.

------. "Russian Politics May Stymie Laser/Optics Collaboration." *Aviation Week* 21 Mar. 1994: 49.

------. "Uzbek Site." *Aviation Week* 15 May 1995: 68-69.

"Secrecy Ends on Spy Agency." *New York Times* 19 Sep. 1992: A5.

Selding, Peter B. de. "Five Launchers Competing for Full-Sized Payloads." *Space News* 8-14 May 1995: 8.

------. "Funding Needed to Ensure Mars Mission." *Space News* 17-23 Apr. 1995: 21.

------. "ILS Unites Atlas, Proton Sales." *Space News* 19-25 Jun. 1995: 9.

------. "Launch Market Prepares for Business War." *Space News* 7-13 Mar. 1994: 8.

------. "Launch Vehicle Firms Predict Market Shift." *Space News* 8-14 May 1995: 8.

------. "Mars Mission Averting German Astronaut's Flight to Mir." *Space News* 27-5 Feb./Mar. 1995: 11.

------. "RKA Chief Blasts Loral's Zenit Deal." *Space News* 19-25 Jun. 1995: 1.

------. "Russia Distances Space Program From Ukraine." *Space News* 20-26 Feb. 1995: 3.

------. "SovCan Star." *Space News* 13-19 Feb. 1995: 1, 21.

Seitz, Patrick. "U.S. Officials Probe Proton, Long March Pricing Policies." *Space News* 7-13 Mar. 1994: 12.

"Shadow Economy." *OMRI Daily Digest* no. 104, pt. 1, 30 May 1995.

Shapiro, Margaret, and Fred Hiatt. "Troops, Yeltsin Foes in Pitched Fight Outside Parliament." 4 Oct. 1993: A1.

Sigel, Thomas. "Economic Decline in 1994." *OMRI Daily Digest* no. 19, pt. 1, 26 Jan. 1995.

------. "Meat and Milk Production Decline." *OMRI Daily Digest* no. 123, pt. 1, 26 Jun. 1995.

"Single-Stage Rocket to Involve Russians." *Space News* 16-22 May 1994: 2.

"Six out of Seven." *Newsweek* 28 Oct. 1957: 35.

"Sixty Warheads Leave Ukraine to be Dismantled in Russia." *New York Times* 6 Mar. 1994: A10.

Socor, Vladimir. "Prime Ministers Highlight Russian-Ukrainian Differences." *Prism* pt. 2, 4 Aug. 1995.

"Soviet, U.S. Scientists Discuss Cooperation in Space." Moscow *Moscow Domestic Service* 18 Nov. 1977. *FBIS* vol. 3: 223; 18 Nov. 1977: 6.

"Space Center Reports on 1994 Activities," Moscow *TASS* 26 Jan. 1995. *FBIS* 30 Jan. 1995: 24.

Specter, Michael. "Climb in Russia's Death Rate Sets Off Population Implosion." *New York Times* 6 Mar. 1994: A1.

------. "New Freedoms Churn a Quiet Russian Village." *New York Times* 14 Aug. 1995: A1.

------. "Russia Promises Budget Curb to Win a Loan of $1.5 Billion." *New York Times* 23 Mar. 1994: A1.

------. "Russians Are Dying Younger and Younger." *New York Times* 2 Aug. 1995: A1.

Spivak, Jonathan. "U.S.-Soviet Cooperation Yields Payload of Down-to-Earth Scientific Benefits." *Wall Street Journal* 18 Jul. 1975: 1, 23.

"Spooked Over Intelligence Cuts." *New York Times* 18 Mar. 1994: A22.

"Sputniks and Budgets." *The New Republic* 14 Oct. 1957: 3-5.

Stanley, Alessandra. "Russia Congress Votes to Release Yeltsin Enemies." *New York Times* 23 Feb. 1994: A1.

Stevenson, Richard W. "Russian Rockets Finding Eager Customers in West." *New York Times* 17 May 1994: D21.

------. "Way Ahead in the Space Race." *New York Times* 5 Apr. 1995: D1.

"Svobodny Upgrades Pegged at $2 Billion." *Space News* 11-17 Jul. 1994: 2.

Sweet, William. "European and U.S. Proposals for Aiding FSU Science Vie for Support." *Physics Today* Jun. 1992: 67-71.

Tarasenko, Maxim V. "Transformation of the Soviet Space Program After the Cold War." *Science and Global Security* 1994, vol. 4: 339-61.

Teltsch, Kathleen. "Space Plans Frustrate the 'Have Nots.'" *New York Times* 14 May 1972: A15.

"Texts of the U.S.-Soviet Agreements on Cooperation in Space Exploration and in Science and Technology." *New York Times* 25 May 1972: A14.

Thornhill, John. "Russian Output Halved in Three Year." *Financial Times* 31 Dec. 1994: 2.

"Twelve Million Russians." *Monitor* vol. 1, no. 54, 18 Jul. 1994.

"Twenty Percent of Russians." *Monitor* vol. 1, no. 63, 31 Jul. 1995.

"Three Killed in Space Center Riot." *Facts on File* 16 Apr. 1992: 266.

"Ukraine Joining Plan for NATO Partnership." *New York Times* 6 Feb. 1994: A8.

"Ukraine Ratifies the Missile Pact, but Delays Ending Nuclear Status." *New York Times* 19 Nov. 1993: A10.

"Ukraine, U.S. Resume Launch Accord Talks." *Space News* 10-16 Jul. 1995: 2.

"The U.S., Ike, and Sputnik." *Newsweek* 28 Oct. 1957: 27-35, 62-64.

"U.S., Russia Settle Export Disagreement." *Aviation Week* 26 Jul. 1993: 27.

"U.S. Said to Life Satellite Restriction." *OMRI* no. 130, pt. 1, 6 Jul. 1995.

"U.S. Team Recommends Cosmodrome Upgrades." *Aviation Week* 28 Mar. 1994: 27.

"U.S.-Ukraine Joint Effort." *Aviation Week* 5 Dec. 1994: 54.

"*V Zerkale Statistiki*" ("Statistics in the Mirror"). *Ekonimika i Zhizn'* (*Economy and Life*) Dec. 1992, no. 51: 1.

"A View from the Cosmodrome." *Washington Times* 10 Mar. 1994: A20.

"Washington Roundup." *Aviation Week* 17 Jul. 1978: 13.

Watson, Traci. "$100 Million Pledged to Support ex-Soviet Science." *Nature* 17 Dec. 1992: 617.

Webb, Jeremy. "Bleak Future for Former Soviet Scientists." *New Scientist* 18 Apr. 1992: 7.

Weldon, Rep. Curt. "Newsmaker Forum." *Space News* 29-4 May/Jun. 1995: 22.

Werner, Debra Polsky. "Aerojet Fires Russian-Made Thruster Rocket in U.S. Tests." *Space News* 23-29 May 1994: 20.

------. "Proton Venture Nears Sellout on Agreement." *Space News* 21-27 Mar. 1994: 3, 20.

"What Next After ASTP?" *Sputnik* (Helsinki, Finland) Apr. 1976: 6-7.

"What's Up? Ask an Auctioneer." *New York Times* 24 Feb. 1994: C1.

Wilford, John Noble. "Soviet Space Papers Going on Sale." *New York Times* 5 Dec. 1993: A36.

Woolsey, R. James. "The Future of Intelligence on the Global Frontier." Address to the Executive Club of Chicago, 19 Nov. 1993.

World Bank Policy Research Department. "Group of Seven Opens New Lending Windows for Russia." *Transition* Apr. 1993.

Yaropolov, Vladimir. "Tombstone Over Zvezdny?" Moscow *Stolitsa* Dec. 1993: 13-15. *JPRS* 3 Mar. 1994: 1.

Yasmann, Victor. "The Russian Election Campaign Begins." *Prism* pt. 1, 21 Jul. 1995.

"Yeltsin Aide: Unemployment to Twenty Percent by Year's End." *Monitor* vol. 1, no. 39, 26 Jun. 1995.

"Yeltsin Issues Edict on Use of Baikonur Cosmodrome." Moscow *TASS* 24 Oct. 1994. *FBIS* 25 Oct. 1994: 25-26.

"Yeltsin Warns." *Monitor* vol. 1, no. 19, 25 May 1995.

Zadorozhnyi, G. P. *"Iskustvennii Sputnik Zemli i Mezhdunarodnoe Pravo"* ("Artificial Satellites and International Law"). *Sovetskaia Rossiia* 17 Oct. 1959.

Zhukov, G. P. "Practical Problems of Space Law." *International Affairs* (Moscow) no. 5, May 1963: 27-30.

------. "Problems of Space Law at the Present Stage." *Memorandum of the Soviet Association of International Law at the Brussels Conference of the International Law Association* Aug. 1962.

------. "Space Espionage Plans and International Law." *International Affairs* (Moscow) no. 10, Oct. 1960: 53-7.

Zolotarev, Viktor. "Russia's Space Policy." *Military Journal #5* (Moscow) no. 70, 1993: 3-23.

Zraket, Charles A. Chair, National Research Council Commission on Physical Sciences, Mathematics and Applications. *Panel to Review EOSDIS Plans: Final Report.* Washington: National Academy Press, 1994.

III. Government Publications

Budget of the United States Fiscal Year. 1995, 1996. Washington: Government Printing Office (GPO), 1994, 1995.

Cheney, Dick. *Annual Report to the President and the Congress.* Washington: Office of the Secretary of the Department of Defense, Jan. 1993.

Commission on Integrated Long-Term Strategy. *Recommended Changes in U.S. Military Space Policies and Programs.* Washington: GPO, Oct. 1988.

Congress and the Nation. 8 vols. Washington: Congressional Quarterly Service, 1965, 1969, 1973, 1977, 1981, 1985, 1989 and 1993.

Department of Defense. "The Soviet Space Challenge." Washington: Office of the Secretary of Defense, Nov. 1987.

Djonovich, Dusan J. *United Nations Resolutions* Series 1, vols. 7-11. Dobbs Ferry: Oceana Publications, 1974 and 1975.

Jasentuliyana, N., and Ralph Chipman. *International Space Programmes and Policies: Proceedings of the Second United Nations Conference on the Exploration and Peaceful Uses of Outer Space.* New York: Elsevier Science Publishing, 1984.

Kecskemeti, Paul. "The Satellite Rocket Vehicle: Political and Psychological Problems." Rand RM-567, 4 Oct. 1950.

Kesaris, Paul. *Documents of the National Security Council 1947-1977.* Washington: University Publications of America, 1980 (microfilm).

Lipson, Leon. "Outer Space and International Law." Rand Paper #P-1434, 24 Feb. 1958.

------. *Legal Problems of Space Exploration: A Symposium* U.S. Senate, 87th Congress, 1st Session (Document no. 87-26). Washington: GPO, 1961.

National Aeronautics and Space Administration. *Aeronautics and Space Report of the President* Fiscal Year 1992, 1993 and 1994 Activities. Washington: NASA, 1993, 1994 and 1995.

President's Science Advisory Committee. *The Space Program in the Post-Apollo Period*. Washington: GPO, Feb. 1967.

Public Papers of the Presidents of the United States: Dwight D. Eisenhower, 1957. Washington: GPO, 1958.

Public Papers of the Presidents of the United States: George Bush, 1992-93. Washington: GPO, 1993.

Public Papers of the Presidents of the United States: John F. Kennedy. Washington: GPO, 1962-1964.

Public Papers of the Presidents of the United States: Ronald Reagan, 1981-1984, 1987. Washington: GPO, 1982-1986, 1989.

Radzanowski, David P., and Stephen J. Garber. "The National Aeronautics and Space Administration." *Congressional Research Service Report for Congress 95-336 SPR*, 1 Mar. 1995.

Report of the Advisory Committee on the Future of the U.S. Space Program. Washington: GPO, Dec. 1990.

Smith, Marcia S. "Russia/U.S. Space Interaction: A Trip Report With Observations and Opinions." *Congressional Research Service Report for Congress 92-774 SPR*. 27 Oct. 1992.

------. "Space Activities of the United States, C.I.S., and Other Launching Countries/Organizations: 1957-1992." *Congressional Research Service Report for Congress 93-379 SPR*. 31 Mar. 1993.

------. "Space Activities of the United States, C.I.S., and Other Launching Countries/Organizations: 1957-1994." *Congressional Research Service Report for Congress 95-873 SPR*. 31 Jul. 1995.

------. "Testimony to the Committee on Science, Space and Technology." *Congressional Research Service Report 6 Oct. 1993*.

Stevenson, Adlai. "Letter to U. Thant." *Department of State Bulletin* vol. 46, no. 1188. 9 Apr. 1962: 588.

The Synthesis Group. *Report to the National Space Council on America's Space Exploration Initiative: America at the Threshold*. Arlington: Synthesis Group, 1991.

Union of Soviet Socialist Republics, United Kingdom of Great Britain and Northern Ireland, United States of America, Afghanistan, Argentina, etc. "Treaty on Principles Governing the Activities of States in the Exploration and Use of Outer Space, Including the Moon and Other Celestial Bodies," signed at Moscow, London and Washington, 27 Jan. 1967 (no. 8843). *United Nations Treaty Series* vol. 610 (1967) 206-19.

United Nations Committee on the Peaceful Uses of Outer Space, 17th Session. "Agenda Item 27: International Co-operation in the Peaceful Uses of Outer Space." *Annexes* Official Record. New York: United Nations, 1963.

United States of America and Union of Soviet Socialist Republics. "Agreement on the Prevention of Incidents On and Over the High Seas," signed at Moscow, 25 May 1972 (no. 12214). *United Nations Treaty Series* vol. 852 (1972) 152-58.

U.S. Congress, Congressional Budget Office. *Setting Space Transportation Policy for the 1990s*. Washington: GPO, Oct. 1986.

U.S. Congress, Joint Economic Committee. *Global Economic and Technical Change* Hearing 16 May and 28 Jun. 1991, S. Hrg. 102-586, pt. 1. Washington: GPO, 1991.

------. *Global Economic and Technical Change: Former Soviet Union and Eastern Europe, China* Hearing 8 Jun. and 27 Jul. 1992, S. Hrg. 102-86, pt. 2. Washington: GPO, 1992.

------. *Measures of Soviet Gross National Product in 1982 Prices* S. Rpt. 101-128. Washington: GPO, 1991.

U.S. Congress, Office of Technology Assessment. *Civilian Space Stations and the U.S. Future in Space.* Washington: GPO, 1984.

------. *Exploring the Moon and Mars: Choices for the Nation.* Washington: GPO, 1991.

------. *The Future of Remote Sensing From Space.* Washington: GPO, 1993.

------. *Launch Options for the Future: A Buyer's Guide.* Washington: GPO, 1988.

------. *UNISPACE '82: A Context for International Cooperation and Competition.* Washington: GPO, 1983.

------. *U.S.-Russian Cooperation in Space.* Washington: GPO, 1994.

------. *U.S.-Soviet Cooperation in Space: A Technical Memorandum.* Washington: GPO, 1985.

U.S. Department of Commerce. *Report of the Advisory Committee on the Future of the U.S. Space Program.* Washington: GPO, Dec. 1990.

------. *Statistical Abstract of the United States* 82nd and 85th eds. Washington: U.S. Bureau of the Census, 1961 and 1964.

U.S. House. Committee on Astronautics and Space Exploration. *The National Space Program* 85th Congress, 2nd Session (H. Rpt. 1758, *Serial Set* 12073). Washington: GPO, 1958.

U.S. House. Committee on Science and Astronautics. *Discussion of Soviet Man in Space Shot* 87th Congress, 1st Session, 13 Apr. 1961. Washington: GPO, 1961.

U.S. House. Subcommittee on International Scientific Cooperation of the Committee on Science, Space and Technology. *International Science and Technology and Foreign Policy* 101st Congress, 2d Session, 4 and 26 Apr. 1990. Washington: GPO, 1990.

U.S. House. Subcommittee on Space Science and Applications of the Committee on Science, Space and Technology. *The Future of the U.S. Space Program* 101st Congress, 2d Session, 23 Jul. 1990. Washington: GPO, 1990.

U.S. Senate. Committee on Aeronautical and Space Sciences. *Documents on International Aspects of Exploration and Use of Space 1954-1962* 88th Congress, 1st Session (Document no. 18, *Serial Set* 12555). Washington: GPO, 1963.

------. *Legal Problems of Space Exploration* 87th Congress, 1st Session (Document no. 26, *Serial Set* 12355). Washington: GPO, 1961.

------. *Soviet Space Programs* 87th Congress, 2d Session (Document no. 87-1). Washington: GPO, 1962.

------. *Soviet Space Programs, 1962-65* 89th Congress, 2d Session (Document no. 89-1). Washington: GPO, 1966.

------. *Soviet Space Programs, 1966-70* 92d Congress, 1st Session (Document no. 92-51). Washington: GPO, 1971.

------. *Soviet Space Programs, 1971* 92d Congress, 1st Session (Document no. 92-2). Washington: GPO, 1972.

------. *Soviet Space Programs, 1971-75* 94th Congress, 1st Session (Document no. 94-2). Washington: GPO, 1976.

U.S. Senate. Committee on Armed Services. *National Defense Authorization Act for Fiscal Year 1993 Report* 102nd Congress, 2d Session (Rpt. 103-352). Washington: GPO, 1992.

U.S. Senate. Committee on Armed Services, Preparedness Investigating Subcommittee. *Inquiry Into Satellite and Missile Programs* 85th Congress, 1st Session. Washington: GPO, 1957.

U.S. Senate. Committee on Commerce, Science and Transportation. *Soviet Space Programs, 1976-80* 97th Congress, 2d Session (Document no. 97-2). Washington: GPO, 1982.

------. *Soviet Space Programs, 1976-80* pt. 2, 98th Congress, 2d Session (Document no. 98-2). Washington: GPO, 1982.

------. *Soviet Space Programs, 1981-87* pt. 1, 100th Congress, 1st Session (Document no. 100-2). Washington: GPO, 1989.

------. *Soviet Space Programs, 1981-87* pt. 2, 101st Congress, 1st Session (Document no. 101-1). Washington: GPO, 1989.

U.S. Senate. Committee on Commerce, Subcommittee on Communications. *The Speeches, Remarks, Press Conferences, and Statements of Senator John F. Kennedy, August 1 through November 7, 1960* 87th Congress, 1st Session (S. Rpt. 994, part 1). Washington: GPO, 1961.

United States Statutes at Large vol. 72, pt. 1. Washington: GPO, 1959.

Vice President's Space Policy Advisory Board. *A Post Cold War Assessment of U.S. Space Policy.* Washington: n.p., Dec. 1992.

Weekly Compilation of Presidential Documents vol. 19, no. 10 (14 Mar. 1983). Washington: GPO, 1983.

IV. NASA History Office

All documents listed in this section of the Works Consulted may be most easily acquired by presenting the bibliographic information presented to one of the historians at the NASA History Office, who organize files variously by subject, time period and persons involved. Each entry is numbered to correspond to references in the text. The NASA History Office is located at 300 E. SW Street, Washington, D.C. 20024, (202) 358-0386.

1. "Agreement Between the United States of American and the Union of Soviet Socialist Republics Concerning Cooperation in the Exploration and Use of Outer Space for peaceful Purposes." 15 Apr. 1987.

2. Bundy, McGeorge. "Memorandum for the President." 13 Jul. 1962.

3. Clinton, William. "Statement by the President." The White House, 17 Jun. 1993.

4. Douglas Aircraft Company, Inc. "Preliminary Design of an Experimental World-Circling Spaceship." Rand Report No. SM-11827, 2 May 1946.

5. Dryden, Hugh L. "Letter to the President." 21 Jan. 1964.

6. Dunbar, Brian. "Successful U.S.-Russian Ozone-Monitoring Mission Appears Over." *News Release* 95-11. 2 Feb. 1995.

7. Fletcher, James. "Statement by Dr. Fletcher." 24 May 1972.

8. Goldin, Daniel S., and Koptev, Yuri N. "Memorandum of Discussion: On Civil Space Cooperation." Moscow, Jul. 1992.

9. Gore, Al. "Remarks by the Vice President in Signing Ceremony with Prime Minister Chernomyrdin of Russia." Office of the Vice President, 2 Sep. 1993.

10. Graham, Wallace H. Letter to Dr. Grosse, 17 Jan. 1973.

11. Grosse, Aristid V. "Report on the Present Status of the Satellite Problem." 25 Aug. 1953.

12. ------. Letter to Dr. Eugene Emme, 12 Jan. 1973.

13. Kennedy, John F. "Memorandum for The Administrator, NASA." 12 Nov. 1962.

14. ------. "Memorandum for the Vice President." 20 Apr. 1961.

15. ------. "Text of Letter Dated March 7, 1962 from President Kennedy to Chairman Khrushchev re Cooperation in Peaceful Uses of Outer Space." 7 Mar. 1962.

16. Khrushchev, Nikita Sergeevich. "Text of Letter from Chairman Khrushchev in Reply to President Kennedy's Letter of March 7, 1962." 20 Mar. 1962.

17. Low, George. "Items of Interest," Memorandum to James Fletcher. 12 Aug. 1971.

18. ------. Letter to the President of the Academy of Sciences of the USSR, 24 Mar. 1975.

19. ------. "Memorandum for the Record: Meeting with the President on January 5, 1972." 12 Jan. 1972.

20. Lunney, Glynn S. "Trip Report - Delegation to Moscow to Discuss Possible Compatibility in Docking." 5 Nov. 1970.

21. Murphree, E. "Memorandum to Reuben Robertson re Jupiter." 5 Jul. 1956.

22. ------. "Use of the Jupiter Re-entry Test Vehicle as a Satellite." 5 Jul. 1956.

23. Murphy, Robert. "Confidential Memorandum for Dr. Alan T. Waterman." 27 Apr. 1955.

24. National Aeronautics and Space Administration Office of International Programs. *International Programs* 31 Jan. 1962.

25. National Aeronautics and Space Administration Office of Life and Microgravity Sciences and Applications. "U.S./Russian Cooperative Human Space Flight Program, Draft Schedule." 1 Nov. 1993.

26. National Security Council Directive 5520. 20 May 1955.

27. "Preliminary U.S. Policy on Outer Space." National Security Council Directive 5814/1. Aug. 1958.

28. Rahn, Debra J. "NASA and Russian Space Agency." *News Release* 94-101. 23 Jun. 1994.

29. Stewart, Homer J. "Memorandum for the Assistant Secretary of Defense (R&D)." 22 Jun. 1956.

30. U.S. Department of State. "Position Paper for U.S. Participation in Legal Subcommittee of UN Committee on the Peaceful Uses of Outer Space." 15 Jul. 1960.

31. "U.S.-USSR Cooperation in Space Press Conference." News Conference in Houston, Texas 13 Jul. 1972.
32. Waterman, Alan T. "Confidential Memorandum for Robert Murphy." 18 Mar. 1955.
33. ------. Letter to the Honorable Donald A. Quarles, 13 May 1955.
34. ------. Letter to Mrs. Constance Green, 15 Jul. 1965.
35. Webb, James E., and McNamara, Robert. "Recommendations for Our National Space Program: Changes, Policies, Goals." May 1961.
36. Zwicky, F. "Report on Certain Phases of War Research in Germany" vol. 1. 1 Oct. 1945.

V. Interviews

Bourke, Roger. Personal Interview. 12 Apr. 1995.

Doty, Paul M. Personal interview. 14 Jan. 1994.

Drovenikov, Igor S. Personal interview. 14 Aug. 1993.

Frutkin, Arnold W. Personal interview. 28 Jan. 1994.

Graham, Loren. Telephone interview. 7 Mar. 1994. Personal interview. 5 Apr. 1994.

Johnson, Nicholas L. Telephone interviews. 8 Mar. 1994, 24 Mar. 1994 and 18 Aug. 1995.

Kurt, Vladimir. Personal interview. 10 Aug. 1993.

Randolph, James. Personal Interview. 11 Apr. 1995.

Rosenberg, Major General Robert A. (USAF, ret). Telephone interview. 25 Mar. 1994.

Smith, Marcia. Telephone interviews. 26 Apr. 1993, 9 Mar. 1994 and 16 Aug. 1995.

Staehle, Robert. Personal Interview. 11 Apr. 1995.

Sokol'sky, Viktor N. Personal interview. 4 Aug. 1993.

Vedeshin, Leonid A. Personal interviews. 4 Aug. 1993 and 13 Aug. 1993.

Index